Taste, Waste and the New Materiality of Food

Anthropocentric thinking produces fractured ecological perspectives that can perpetuate destructive, wasteful behaviours. Learning to recognise the entangled nature of our everyday relationships with food can encourage ethical ecological thinking and lay the foundations for more sustainable lifestyles.

This book analyses ethnographic data gathered from participants in Alternative Food Networks from farmers' markets to community gardens, agricultural shows and food redistribution services. Drawing on theoretical insights from political ecology, eco-feminism, ecological humanities, human geography and critical food studies, the author demonstrates the sticky and enduring nature of anthropocentric discourses. Chapters in this book experiment with alternative grammars to support and amplify ecologically attuned practices of human and more-than-human togetherness. In times of increasing climate variability, this book calls for alternative ontologies and world-making practices centred on food which encourage agility and adaptability and are shown to be enacted through playful tinkering guided by an ethic of convivial dignity.

This innovative book offers a valuable insight into food networks and sustainability which will be useful core reading for courses focusing on critical food studies, food ecology and environmental studies.

Bethaney Turner is an Assistant Professor in Faculty of Arts and Design at the University of Canberra, Australia. Her interdisciplinary research explores how more sustainable urban living behaviours can be developed and fostered in a time of human-induced climate change.

Critical Food Studies

Series editor: Michael K. Goodman
University of Reading, UK

The study of food has seldom been more pressing or prescient. From the intensifying globalisation of food, a world-wide food crisis and the continuing inequalities of its production and consumption, to food's exploding media presence and its growing re-connections to places and people through 'alternative food movements', this series promotes critical explorations of contemporary food cultures and politics. Building on previous but disparate scholarship, its overall aims are to develop innovative and theoretical lenses and empirical material in order to contribute to—but also begin to more fully delineate—the confines and confluences of an agenda of critical food research and writing.

Of particular concern are original theoretical and empirical treatments of the materialisations of food politics, meanings and representations, the shifting political economies and ecologies of food production and consumption and the growing transgressions between alternative and corporatist food networks.

Practising Empowerment in Post-Apartheid South Africa
Wine, Ethics and Development
Agatha Herman

Digital Food Activism
Edited by Tanja Schneider, Karin Eli, Catherine Dolan and Stanley Ulijaszek

Children, Food and Nature
Organising Meals in Schools
Mara Miele and Monica Truninger

Taste, Waste and the New Materiality of Food
Bethaney Turner

For a full list of titles in this series, please visit: www.routledge.com/Critical-Food-Studies/book-series/CFS

Taste, Waste and the New Materiality of Food

Bethaney Turner

Routledge
Taylor & Francis Group
LONDON AND NEW YORK

First published 2019
by Routledge
2 Park Square, Milton Park, Abingdon, Oxon OX14 4RN

and by Routledge
52 Vanderbilt Avenue, New York, NY 10017

Routledge is an imprint of the Taylor & Francis Group, an informa business

British Library Cataloguing-in-Publication Data
A catalogue record for this book is available from the British Library

Library of Congress Cataloging-in-Publication Data
Names: Turner, Bethaney, author.
Title: Taste, waste and the new materiality of food / Bethaney Turner.
Description: Abingdon, Oxon ; New York, NY : Routledge, 2018. |
Series: Critical food studies series | Includes bibliographical references and index.
Identifiers: LCCN 2018032994| ISBN 9781472487544 (hbk : alk. paper) |
ISBN 9780429424502 (ebk) | ISBN 9780429755187 (mobi/kindle) |
ISBN 9780429755200 (pdf) | ISBN 9780429755194 (epub)
Subjects: LCSH: Food consumption–Moral and ethical aspects. |
Food consumption–Environmental aspects. | Taste. | Fair trade foods. |
Food waste.
Classification: LCC HD9000.5 .T83 2018 | DDC 178–dc23
LC record available at https://lccn.loc.gov/2018032994

ISBN: 978-1-472-48754-4 (hbk)
ISBN: 978-0-367-58307-1 (pbk)
ISBN: 978-0-429-42450-2 (ebk)

DOI: 10.4324/9780429424502

Typeset in Times New Roman
by Integra Software Services Pvt. Ltd.

Contents

Acknowledgements

This book would not have been possible without the generosity of the gardeners, agricultural show exhibitors and Alternative Food Network shoppers who invited me and my fellow researchers into their homes and gardens. The participants' willingness to share their time, produce and stories with me made this research such a joyful experience. My colleagues Joanna Henryks, Cathy Hope, George Main and Kirsten Wehner played key roles in some of the research projects that fuelled the ideas and arguments developed throughout this text. I extend my thanks to them for stimulating my thinking. I also need to express my gratitude to my very efficient Research Assistants, Nic Mikhailovich and Li Nguyen, who worked diligently on these projects at times when I was swamped by other commitments. Thanks must also go to Sue North and Nick Potter for their willingness to cast a critical eye over the work. Mike Goodman has been a patient and supportive series editor and it is a great honour to be able to contribute to the body of work developed in the Critical Food Studies series. Small sections of this research have been published previously and I thank the publishers for granting permission to reprint this here.

Chapter 6 includes sections originally published in: Bethaney Turner, Joanna Henryks, George Main & Kirsten Wehner (2017) Tinkering at the Limits: agricultural shows, small-scale producers and ecological connections, *Australian Geographer*, 48:2, 185-202. copyright © Geographical Society of New South Wales Inc., reprinted by permission of Taylor & Francis Ltd, http://www.tandfonline.com on behalf of Geographical Society of New South Wales Inc.

Chapter 9 includes sections originally published in: Bethaney Turner (2018). Playing with food waste: Experimenting with ethical entanglements in the Anthropocene. *Policy Futures in Education*, (Online only at this stage) Copyright © [2018] (SAGE Publications Ltd). Reprinted by permission of SAGE Publications. DOI: 10.1177/1478210318776851

1 Introduction

Introduction

Catastrophic visions of the future abound in scientific reports addressing climate change. Food is often a central focus of these dystopic predictions, with fears raised about our ongoing capacity to produce sufficient food in the wake of increasing climate variability and the accompanying vulnerabilities. This uncertainty bleeds into the present where extreme weather and its impacts—from lost crops to skyrocketing food prices—are increasingly experienced. Despite this, global responses are slow and unwieldy. Recognition of human induced climate change has led to calls to name this geological age the Anthropocene (Steffen, Crutzen & McNeill, 2007; Steffen, Grinevald, Crutzen & McNeill, 2011; Steffen et al., 2011), designating it as the first era in which humans rather than 'natural' forces have had the most significant impact on the planet. The debates about terminology are as lively as those concerned with just how our planetary futures may look. However we choose to name it, model it and represent it, the climate is changing and we humans—specifically those of us in the minority world—have been at the forefront of instigating these shifts. On a daily basis we will have to take steps to mitigate our ongoing impact and adapt to the changes already in train through shifts in what we eat as well as how and where we produce our food. This means that, for many of us, the future will have a different taste.

But, this is not a book about the science of climate change. Nor is it one that dwells on doomsday scenarios, though such dire imaginings are a haunting presence throughout. Instead, it is a book about how people engage with food in a time of climate variability and how particular configurations of human/more-than-human relations in the production, distribution, ingestion and wasting of food might help lay the foundations for more sustainable futures. These futures rely on reconceptualisations of anthropocentric forms of subjectivities and narratives, and so my primary concern in this text is ontological in nature. It is about how we can be and do in the world in ways that acknowledge and are responsive to our mutual entanglements. The centrality of food in our daily lives makes it a productive optic through which we can highlight both the destructive modes of being and doing that dominate the Anthropocene as well as the

DOI: 10.4324/9780429424502-1

possibilities for enactment of more hopeful alternatives. The skills and capacities to do things differently are enlivened in this book through stories of everyday people and their relations with the more-than-human world that revolve around food, from gardens to garbage tips, from production to decomposition, from taste to waste. Drawing on these food stories we can sense alternative ways of being in—or more precisely becoming-with—the world that underpin less resource intensive lifestyles and which may provide glimpses of how humans and more-than-humans can live well together through attunement to mutual vulnerabilities, willingness to enter into risky encounters and via responsiveness to the affective force of our relational entanglements. While I stretch these stories beyond their geographical and temporal locations, they are very much grounded in particular places and their unique affordances.

Food in the city: exploring ethico-political potentialities

The food stories in this book unfold in the specific context of the urban and suburban realm of Canberra, Australia's capital city, but in them I hear echoes from elsewhere, highlighting the possibility of bigger, broader changes in minority world cities across the globe. Cities demand our attention in the Anthropocene. Fuelled by the resource-intensive machinations of industrialisation, they are now home to more than half the world's population. Until recently, urban planning and development has paid little attention to food and even less to the qualities of urbanites' relationships with the food system. Yet, food and cities are melded together in a variety of deep and shifting ways (Turner & Hope, 2015). Industrial agriculture has both shaped the development of cities, and been shaped itself, in ways that respond to the various forms that the urban explosion has taken. We see this particularly in the practices of food cultivation and transportation that have become so dominant in the modern urban context (Steel, 2009). These shifts, perhaps most obvious in the distancing of food production from cities, underpin a rural/urban divide commonly expressed as a nature/culture binary that permeates imaginings of the city in the minority world (Barthel & Isendahl, 2013; Jarosz, 2008; Pothukuchi, & Kaufman, 1999; Sonnino, 2009). The distancing of food growing from cities is said to contribute to a 'denaturalisation of urban imaginaries' (Blecha & Leitner, 2014, p. 87). City infrastructure and transportation routes, as Carolyn Steel notes, have 'emancipated cities from geography' (Steel, 2009, p. 38), removing the restrictions of the particular climatic and ecological conditions in which they are built by enabling food to flow across vast global distances. In so doing, discursive and practical renderings of the city as human-constructed or orchestrated spaces and places ('built environments') have tended to promote understandings of ecological conditions as passive, surmountable and, perhaps most disconcertingly, distinctly separate to the human lifeworld (Blecha & Leitner, 2014; Pincetl, 2007; Steel, 2009).

While representations of a human/nature divide dominate conceptualisations of cities, such understandings fail to attend to the everyday practices of some urban dwellers and the materiality of their ecological and food encounters. So, while we should indeed work to resist dominant anthropocentric understandings, we also need more generative approaches capable of attending to the complexity of urban relations and their productive political potential. To explore some of the specificities of how this can play out, I draw on Hinchcliffe and Whatmore's (2006) notion of living cities, which they claim are brought to life as assemblages amongst human and nonhumans. Recognition of cities as 'lively' requires engagement in 'a politics of conviviality that is serious about the heterogeneous company and messy business of living together' with Hinchcliffe and Whatmore noting:

> The notion of living cities fleshes out a sense of ecological co-fabrication in which the life patterns and rhythms of people and other city dwellers are entangled with and against the grain of expert designs and blueprints. This conceptual shift from built environments (as they are termed in conventional Town and Country Planning) to living cities is allied to a realignment of the politics of nature such that cities are appreciated as 'ecological disturbance regimes rather than ecological sacrifice zones' (Wolch, 1998) in which people are no longer considered inimical to nature, nor natures antithetical to cities.
> (2006, p. 134)

It is through attunement and responsiveness to this very messiness of living in a space of 'ecological co-fabrication' (Hinchliffe & Whatmore, 2006, p. 134) that simplistic urban/rural, nature/culture divides break down enabling creative reconfigurations of human/more-than-human relations to become possible.

The inescapability of 'eating', and thus the necessity of engagement with the food system, provides sites ripe for exploration of how these forms of responsiveness can play out and throughout this book we will encounter examples of creative reconfigurations in urban food practices. These reconfigurations are, invariably, marked by openness to the multiplicity of assemblages in which the necessity of 'togetherness' comes to the fore. For some this is expressed through forms of attachment or connection incited by recognition that '[w]e are all just different collections of the same stuff—bacteria, heavy metals, atoms, matter-energy—not separate kinds of being susceptible to ranking' (Gibson-Graham, 2011, p. 3). But, these are rarely harmonious, comfortable relations. For others, togetherness can also be marked by acknowledgement of shared alterity, sometimes manifesting in detachment, but always cognisant of mutual vulnerabilities often induced by acknowledgement of our necessary 'togetherness' in the world (Abrahamsson & Bertoni, 2014). Such forms of living together in urban sites offer hope—not in an unfounded optimistic sense, but in an attuned, responsive fashion that must be continuously worked on (Head, 2016)—for the development, or enhancement, of practices that challenge the hyper-separation of human exceptionalism that is the hallmark of anthropocentric thinking and so commonly represented as defining urban metropolises.

To explore these big-picture urban challenges through the lens of the everyday, this book focuses on fine-grained analysis of the alternative food related practices in one city, Canberra. This micro-level focus helps us to reconsider dominant narratives and identify alternative practices that, while small in scale, may be ripe for amplification. It is here in the daily lives of city inhabitants and their food stories that the stirrings of new ways of doing, being and becoming together may well be sensed. Before we dive into these food stories, let me contextualise them within their geographical and ecological co-fabrications; let me introduce you to the city in which they live.

Encountering food in Canberra, Australia

Canberra, constructed in 1913, is an entirely planned city designed by the winner of an international competition, the American architect, Walter Burley Griffin in partnership with his wife, Marion Mahoney Griffin. The 'bush capital', as it is known, lies on the vast limestone plains of Ngunnawal country and is designed to open itself up to the natural environment, to be 'of' not just 'in' the landscape (Taylor, 2006). It is divided into four key town centres with a more-or-less central metropolitan area known as Civic. Between these town centres are swathes of bush land—scrubby, dry sclerophyll forest most notably inhabited by eucalyptus trees (known here as widow-makers for their propensity to unexpectedly drop their weighty limbs). The city is planned around the central parliamentary triangle composed of two mountains (hills by most other nations' standards) and Parliament House. While the building that housed our first governments is now a museum, a larger parliamentary building is located directly to its rear, maintaining the planned symmetries as best as possible. New Parliament House, is built 'into' the soil, part of its sloping roof covered in lawn, a delight for visiting school children to roll down. Being able to walk on our House of Parliament is supposed to be a reminder that parliamentarians are delegates of us Australians, though, as I write this, rising concerns of terror threats have prompted a rethinking of this largely unfettered roof access. While the development of Canberra has continued to promote and support the 'bush capital' ideal, other components of the city's planning have largely fallen to the wayside, notably the designers' original interest in local food production.

In the city's early days, local food (AGHS, 2006) production was viewed as critical to a thriving city. In fact, at the time, the Federal Government was keen for Canberra to become self-sufficient. To realise this, the fertile alluvial soils of Pialligo on the city's outskirts were earmarked for commercial-scale vegetable production (Turner, Pearson & Dyball, 2012). Pialligo retains elements of its rural character today, though large-scale quantities of food are not produced here and the recently expanded airport is on its doorstep. Private householder backyards were also considered to play a key role in realising self-sufficiency. The 'Garden City' ideals of Ebenezer Howard permeated the subdivision practices employed from the inner city to the suburbs. Canberra house plots were to be large enough for homeowners to have backyard vegetable gardens, chickens and fruit trees (AGHS, 2006) as well as utility areas (sheds and clothes lines) and some lawn for recreation (typically, enough to bowl a cricket ball). However, today, commercial buildings are rooted in much of the city's most fertile soils.

The current planning focus on urban in-fill and the shrinking footprint of urban backyards in greenfield sites due to reduced block sizes and larger houses have significantly altered Canberrans' potential encounters with food production and local food. There is no longer a local dairy, and while livestock is raised for meat there is no abattoir to process it. Food production has been gradually removed from the city (Turner & Hope, 2015; Turner, Pearson & Dyball, 2014). Canberra may be designed to be 'of' its natural environment, but the capacity of this landscape to produce food and nourish alternative conceptions of the city remains largely silenced (Turner & Hope, 2015).

In recent years we have seen a global backlash against this sort of distancing of food from city centres with a veritable explosion of urban agricultural practices across the minority world, from community gardens to farmers' markets. This is true also of the Canberra experience where the last 15 years have seen the introduction of two weekly farmers' markets, the expansion of community gardens from three in the year 2000 to at least 17 in 2016, and the introduction of two farmers' retail outlet shops. In later chapters, we will meet a range of everyday people—gardeners, farmers' market shoppers, composters—and explore the ways in which they engage with these initiatives, and food more broadly, in a time of climate change. Theirs are stories permeated by sensorial engagements. The fieldwork was inevitably punctuated by the smells, tastes and textures of garden produce and compost. I never left a food-producing garden without either having been fed or gifted with armfuls of newly harvested goodies. In all the transcripts where I still have the recorder running, my protestations are evident—'No, you have already given me so much of your time. I can't possibly take this (as delicious as it looks). Surely you can use it?' This never worked though. I left with armfuls of food for which I was very grateful. As one gardener said to me, after I had followed him around chatting as he watered his newly planted winter crops at dawn on a day threatening to push into the mid-30s, 'Look, I think you can't come to the garden without me giving you something.' As I took hold of a bag full of the last of his summer crops of tomatoes and cucumbers, he smiled and said, 'Yeah. Just be careful. Most of them are pretty grub free, but there might be the occasional one.'

In the following pages, I attempt to explore how the everyday actions of the gardeners, farmers' market shoppers, agricultural show exhibitors and composters who participated in this research enact ways of being and doing in the world that differ from contemporary representations of profligate, unthinking, unfeeling consumers acting in accordance with a doctrine of human exceptionalism. Those who gave their time to this research are, with a few notable exceptions, people who do not identify as activists, many of them explicitly stating that they were not 'greenies' or 'hippies' (even when no questions were asked about their environmental politics). This identification, or lack-thereof, aligns closely with Head's observations that living less resource intensive lifestyles does not need to be underpinned by a 'green' ethic (Head, 2016, pp. 167–173).

I do not attempt to coalesce these people into new subjectivities, but I dwell with their stories and practices and invite you to do likewise, to learn with them. I do see in their practices tensions between a hopeful future and one where they are overwhelmed by its very uncertainty. I see expressions and evidence of attachments and detachments to 'big picture' issues such as climate change through to micro-level concerns such as shifts in their soil conditions. Most overwhelmingly, I find in their everyday actions a form of 'togetherness' with the more-than-human world. This is not always harmonious and such forms of being or becoming-with can sometimes be silenced when more immediate concerns press on their minds and bodies. But, by and large, their engagements with food represent forms of attunement to our relational entanglements that are glimpsed in the quotidian experiences of the everyday; where they shop, how they prepare and consume their food and what becomes of the waste they generate.

Looking here at the minute detail of how we live inevitably elicits rousing calls of 'this can't be enough'. 'What can such small-scale action do in the face of significant climatic changes and uncertainty?' I do not intend to dismiss these concerns, nor deride them. I share the sentiments. I too want more. I think we can do more. Our governments, industry, and international organisations are failing us on a daily basis when they give up on opportunities to effect more sustainable changes to the ways our lives play out (particularly for those in the minority world). But, I also think that policy, quotas and taxes are not enough. The perpetuation of anthropocentric attitudes is a key barrier to more sustainable change. While little at the policy level seems to challenge these beliefs, the very notions of human exceptionalism and hyper-separation are regularly questioned in the lives of those we meet 'on the ground' in this book. These confrontations with the lack of our human control are not always welcomed by the participants and can be actively resisted, but it is something that tends to become 'accepted'. Knowing we cannot disentangle ourselves from the multiplicity of more-than-humans that make our lives possible doesn't move us very far, but 'being' it and 'doing' it just might. In their everyday 'beings' and 'doings', we see how these people actually live differently—responsively and response-ably—with the more-than-human world. I couch these 'beings' and 'doings' within ontological and theoretical arguments, but above all they are embedded in embodied, visceral 'real' world actions playing out in the everyday lives of those involved in this research. It is here in these daily actions and our capacity to amplify these through careful grammars of action that hope for more sustainable futures may well take root. To develop these lines of argument throughout the book, I draw on work from a number of fields that encourage attunement to the material world and contribute to the development of responsive forms of ethico-political practices and narratives. So, just for the moment, please excuse me as I move away from food to sketch out the bare bones of this theoretical terrain.

Sketching out the theoretical terrain: attuning to material relations

This book draws on key concepts that I see as complementary across fields such as geography, political ecology, eco-feminism and cultural theory to name a few. At its core, it builds on strands of arguments developed across these disciplines that call for attention to be drawn away from the individual human actor to highlight the materialities of the world. As Whatmore (2006) notes, this so-called 'material turn' is not new nor homogenous, but the context in which we encounter it may well be. Climate change is now palpable, visceral and affective in force. It has been induced by anthropocentric material and discursive practices that hail a world predicated on the hyper-separation of the human from what is commonly represented as 'nature'. This old 'mode of humanity' (Plumwood, 2001) centred on perceptions of human exceptionalism and perpetuation of Cartesian notions of an individual subject capable of unilaterally acting in, and on, the world with free will has failed to secure a safe future for the planet and its inhabitants. It is increasingly apparent that we need new ways of being and becoming, attuned to the very materialities of our world if a survivable future is possible. These attunements may well allow us to acknowledge and live with both the grief induced by losing familiar modes of living and with hope for the future (Head, Atchison & Phillips, 2015)

There is no blueprint for just how these different ways of living and being will look, nor how we cultivate them. However, along with many encountering materiality in forceful ways (Gibson-Graham, 2008, 2011, 2015; Harbers, Mol & Stollmeyer, 2002; Head et al., 2015; Heuts & Mol, 2013; Law & Mol, 2008; Mol, 2008; Mol & Law, 2004), I argue that these will only arise through experimental means. We have to be ready and willing to do things differently, and be open to surprise and the unexpected. Doing this will likely involve cultivation of forms of what Haraway designates as response-ability, 'a relationship crafted in intra-action through which entities, subjects and objects, come into being' (Haraway, 2008, p. 71), and an openness and sensitivity to Latour's notion 'to learn to be affected, meaning "effectuated", moved, put into motion by other entities, humans or nonhumans' (Latour, 2004, p. 205). The ground will be constantly shifting as Law and Mol (2002, p. 20) write, 'Multiplicity, oscillation, mediation, material heterogeneity, performativity, interference. . .there is no resting place in a multiple and partially connected world'.

The most productive experimentations are not likely to involve humans finding out about the world and acting on it anew, but ones that take seriously more-than-humans as active, relational participants in this process. This is configured in multiple ways in the existing literature loosely referred to as new materialism and anchored in forms of post-humanist thinking. Jane Bennett's notion of 'thing-power', 'a materiality experienced as a lively force with agentic capacity' which 'could animate a more ecologically sustainable public' (2010, p. 51) is one of the most well-known of these. Bennett conceptualises this idea through her notion of 'lively matter' and the 'thing power' of objects which she defines as 'the curious ability of inanimate things to animate, to act, to produce effects dramatic and subtle' (Bennett, 2010, p. 351). For new materialists, if we are willing to read agency into the nonhuman things around us, then we become forced to recognise (or attuned to the realities) that humans are simply one more element of a world of things that can act on, with or against others through various assemblages. These assemblages can be made, undone and rebuilt in multiple ways. The power of the elements to act within these may not be equal, but nor are they stable, static or predictable. We do not know what relations will form and what their outcomes may be. Our engagement in these is necessarily experimental.

For Bennett, and others occupying the new materialist landscape, this is not simply a return to previous materialist theories premised on naïve notions of object agency. It is instead an approach motivated by attempts to develop understandings and strategies that encourage engaged ecological living practices that seek to avoid ongoing human-inflicted environmental damage caused by the 'master rationality' (Plumwood, 2001) that fuels the Anthropocene. Others who also take materiality seriously are often bundled under the term 'new materialism', but one can sense some authors bristle at the term. Those enacting material feminism emphasise 'engagements with matter' (Hird, 2014, p. 330), including Haraway's focus on companion-species making and multispecies flourishing through her natureculture (material-semiotic) approach (Haraway, 2008) and Hird's micro-ontology that seeks to engage with the microcosmos where most of those encountered are not species but bacteria which 'not only evolved all life (reproduction, photosynthesis and movements) on Earth; they provided the environment in which different kinds of living organisms can exist' (Hird, 2010, p. 37).

Those concerned with the material stress that attention to materiality isn't new nor singular (Whatmore, 2006). Others emphasise that attunement to materiality necessitates a focus on relations, not just the capacities of the singular actors involved, referring to this as matter-in-relation (Abrahamsson, Bertoni, Mol & Ibáñez Martín, 2015) or relational materiality (Law, 2009). Regardless of differences in theoretical positioning, all of those working with the material speak of the shifting, messy entanglements of human and more-than-humans. They speak of the need to engage in experimentation that enables reciprocal attuning or adjusting of ourselves with matter. As Paulson, drawing on Stengers, writes:

> ...it is not enough to decide to include nonhumans in collectives or to acknowledge that societies live in a physical and biological world, as useful as these steps may be. The crucial point is to learn how new types of encounter (and conviviality) with nonhumans...can give rise to new modes of relation with humans, i.e., to new political practices.
>
> (Paulson, 2001, p. 98)

To take this one step further, the concern is not only political but also ontological. Not only do we need experiments in living, but experiments in being.

Playful tinkering with new worlds: developing practices and grammars of risk and hope

Adjusting to the need for experimentations, this text develops the concept of play enacted within an ethico-political framework and alternative grammar articulated as convivial dignity which attempts to offer a guide to experiments in a becoming world that are marked by hope and punctuated by 'an affect of uncertain excitement' (Gibson-Graham, 2011, p. 3). The concept of convivial dignity is itself experimental, arising in response to challenges in conceptualising practices encountered in the case studies. As such, play enacted with convivial dignity is a concept offered in the spirit of Actor-Network Theory (ANT) (though the arguments herein do not necessarily draw only on this approach) insofar as it draws on and employs the open-endedness of ANT to contribute to the shaping of a conceptual framework aimed at inciting productive ways of developing and understanding new modes of being and doing in the world.

Play is offered as a particularly malleable variant of experimentation, far removed from discussions of the role of experts, able to be enacted by all and focused on everyday actions. It is also suggested to be something that develops in a tinkering mode, that unfolds 'bit by bit' in a manner that adjusts and responds to others involved in the encounter (Mol, Moser & Pols, 2010, p. 14). Convivial dignity itself plays with words and meaning in its efforts to prod, probe and prompt new non-anthropocentric grammars; as Mol writes of ANT 'over the years, again and again, new words have been borrowed, invented, adapted. They open new possibilities and throw up surprising insights' (2010, p. 264). While this may appear to be just another neologism, my intention here is to encourage debate about the paucity of grammars we have to conceptualise ways of being and doing in the world that challenge human exceptionalism. These challenges are responsive to the embodied, visceral mode of these encounters where words are not enough, but I contend throughout that a focus on the material cannot come at the expense of representation. We need new ways of talking about these practices because these may well help open up greater possibilities of enacting them.

Along with Gibson-Graham, who observe 'that a discursive politics, or politics of language, [has] had a role to play in any collective action' (2014, p. 81), the argument in this text is grounded in the belief that the language we use is vitally important. There is a dearth of words and grammars to support post-humanist conceptions attuned to materialities that seem necessary to support experimentation with, and amplification of, ways of being and doing that are sensitive to the more-than-human world and imbued with an understanding of the need to include an uncertain future in our present lifestyle choices. Yes, this is partly an outcome of the 'more-than-we-can-tell' (Carolan, 2016, p. 147) embodied, sensorial, affective encounters with food, but I am interested here in how we can represent these in ways that are not readily dismissed. This, of course, is congruent with the ideas of those working with embodied, visceral understandings. As such, I draw heavily on this work as I seek to highlight the narrative limitations impeding expression of encounters with human/more-than-human assemblages and offer the experimental notion of playful tinkering enacted with convivial dignity as one way of broadening our ecologically and viscerally attuned grammars.

The ethico-political demands of convivial dignity: forging forms of togetherness-in-relation

Convivial dignity is discussed here as a guiding ethico-political framework for playful practices attuned to, and primed for, surprise. As such, it is not a fixed, stable entity. Like play itself, it is relational in an active, ongoing, unfolding sense. Convivial dignity offers a touchstone for an ethics of playful tinkering in the Anthropocene. Dignity is used as an intentional provocation to the privileging of the species scale in the term's usual deployment in rights and ethical discourse. The Anthropocene, as Head (2016) points out, highlights the limitations of a species-level approach. Not all humans have acted equally to induce climate change, and the concept of the Anthropos runs the risk of perpetuating humanist assumptions of human exceptionalism. Convivial dignity subverts the species as the key marker of dignity, stretching it to incorporate the entangled relational engagements—and most importantly the relations themselves—with more-than-humans. After all, every species, Haraway reminds us, 'is a multi-species crowd' (2008, p. 165). Convivial dignity extends this beyond species to the microcosmos of bacteria (Hird, 2014), non-living entities and the relational flows through which these shift and congeal in various forms.

The term convivial is used to emphasise that the realisation of this reformulation of dignity is a necessarily co-operative endeavour, enabled only through our relational entanglements. It echoes the 'politics of conviviality' espoused by Hinchliffe and Whatmore which they state 'is serious about the heterogeneous company and messy business of living together' (2006, p. 134). So, convivial dignity is both ethical and political in its drive. It is offered as a way of supporting new ecological narratives that emphasise connections and multispecies becomings. It is a concept that both enacts, as well as requires others to enact it in our task of learning how to be and do differently with the world. It offers a touchstone for forms of what I call togetherness-in-relation. This is togetherness that is not reliant on connections that can be conceptualised as attachments or detachments, nor is it something necessarily mobilised by calls or desires to 'care', yet it is has significant affective force. The togetherness-in-relation supported by convivial dignity is far more open. It moves beyond human exceptionalism even when this may well be desired by the humans involved. The very limits of anthropocentric ways of being and doing are experienced by those involved in the fieldwork for this research in their everyday encounters with togetherness-in-relation

This experimentation with convivial dignity is risky. The term risks being misrecognised as tied to modern notions of the human subject. Building on this, it is also in danger of being misinterpreted as a form of essentialism due to the existing positioning of dignity as an inalienable human right common to all at this species level. Finally, calls for dignity typically emanate from the disempowered—those who perceive that their dignity has been violated. In this work, I am very cautious about claiming to speak for others—both humans and more-than-humans. However, risks need to be taken if we are to encourage a less resource intensive, more attuned planetary existence. Stengers sees risk as a 'concrete experience of hope for change' (2002, p. 246), and it is in this vein that the term is employed here. Though, throughout this text I will suggest that engagement in risk is likely to be more generative if it is recast as a condition of playful tinkering enacted with convivial dignity. Throughout all of these discussions I actively attempt to avoid positioning any of the human voices here as benevolent subjects allowing the 'other' to speak or be heard. These challenges plague all of us working with the material, and work in these spaces requires ongoing diligence to heighten sensitivities to materialities. But it is exactly this question of how can we heighten and amplify these sensitivities to support new ways of being and doing in the world that we must examine. This aim underpins the ideas developed here and makes unique demands of our methodological approaches.

Multisensorial encounters: the necessity of engaging with talk-in-relation

Multisensorial engagements with the material world demand that we expand our methodological approaches to incorporate methods less reliant on direct human input. This includes the use of automated cameras (Pitt, 2015), GoPros, drones and audio-recordings/soundscapes to name a few;these hold great potential. This text, however, primarily engages with human action, perception and talk recorded in diaries, photos, surveys, participant observation, informal conversations and formal interviews related to food practices. Echoing Anderson's oral history aims, I focus on enabling people to share their stories 'in their own words' while being immersed in their everyday food experiences (Anderson, 2014, p. 58). In these encounters, those sharing their stories often draw on a number of competing discourses to make sense of their interactions with food. On the one hand, the stories are riddled with recognition of more-than-human vitality; on the other, they often indicate a desire to exert control over these vitalities. This is also evident when the 'telling' of the stories is replaced with showing ('see look here', 'see how this grows' or 'smell this') as participants dwell in their gardens, exhibition halls, farmers' markets and kitchens. In these encounters, showing is always accompanied by doing—pulling weeds, picking off bugs, harvesting produce, smelling, touching and turning the compost to name a few. In the doing and showing, the embodied sensorial practices highlight that words alone are not enough to capture these everyday food interactions. As Law & Mol write, 'while material politics may well involve words, it is not discursive in kind' (2008, p. 133). The viscerality, embodied, multisensorial nature of our daily encounters with food makes it a fruitful focus for research concerned with how we can live differently with the world in a time of anthropogenic climate change.

Subjectivities, matter and food: an invitation to play in the Anthropocene

My focus on food takes sustenance from Mol's suggestion that to conceive of subjects anew in the Anthropocene we need to 'play with our food, that is, explore the possibilities of models to do with growing, cooking, tasting and digesting' (2008, p. 34). Food also enables us to keep the discursive in view through efforts to attend to what Goodman calls the 'grammars of eating' to 'further situate the relationalities of food's vital materialisms and its shifting geographies' (Goodman, 2016, p. 264). As such, this book is an invitation to play. At times this becomes a shrill plea. I try to differentiate this plea to play from that of the 'bored' child seeking attention, but in some ways it echoes the desire common to children to enter relations with others where new ways of being and doing can be imagined and made possible. The play I invite you to join here revolves around experimenting with how humans and more-than-humans can be and 'do' together in ways that resist human exceptionalism.

I am interested in how attunement to our entangled lives revolving around food can support more sustainable futures by simultaneously drawing attention to destructive impacts of anthropocentric practices and its narrative power, while simultaneously pointing to the presence of alternative practices in the lived realities of urban residents. The book aims to not only speak 'to a subject who is prepared to include the insecure future within a careful present' (Mol, 2008, p. 29), but to attempt to locate some moments of practice and representational strategies that help spread and promote this. The outcome will not be a one-size-fits-all universal model but a series of prompts for ongoing playful tinkerings with doings in diverse contexts. As such, it aims to respond to Gibson-Graham's question, 'How might new forms of social and ecological connection arise?' (2011, p. 12) by acknowledging that, as Hinchliffe and Whatmore note,

> ... human bodies are set up or configured in the world in particular fashions, the fashioning of those worlds can amplify or otherwise these configurations. The attachments are therefore part of a mutual pushing that can produce a 'feeling life (in the doubled sense of both a grasp of life, and emotional attunement to it)'.
>
> (2006, p. 133)

The task then is to play with the attachments and, as we will see, also with forms of detachment and more multifarious forms of togetherness that might enable amplification of a 'feeling life' attuned to the more-than-human entangled becomings that make life, and our futures, possible.

To both mitigate and live with the damage of anthropogenic climate-change, we will need to engage in more agile, responsive forms of living and being. Anthropocentric narratives and practices need to be resisted. Cities, the hallmark of human socio-technical power, provide a site where this resistance is needed and also a place where future vulnerabilities will be keenly felt as more pressure is placed on existing infrastructure. These are also sites where new possibilities are emerging, where recognition of mutual vulnerabilities can encourage engagement in risky encounters, and where I find some inhabitants are willing to playfully tinker with new forms of responsive interactions with the more-than-human. But just how do we make these matters-in-relation matter? I respond to these questions by following Gibson-Graham's calls to 'start where we are' (2008, p. 614) and Tsing's assertion that 'Familiar places are the beginning of appreciation for multi-species interactions' (2012, p. 142). I start then with everyday encounters with food in the city I call home, Canberra.

In the food stories explored here I contend that we can identify alternative ways of being, doing and becoming-with through the training of sensitivities to matter-in-relation that enables us to be moved by the affective force of our entangled human/more-than-human relations. I argue that these are enacted through the generative practice of playful tinkering and point to the need for new grammars that can assist us to support and amplify the potential of the alternative forms of subjectivity enacted here. While the power of anthropocentric narratives and the resistance of the material to narrativisation are challenges, I attempt to contribute to efforts to promote alternatives through the development of the ethico-political concept of convivial dignity. Convivial dignity is offered in the spirit of an experiment as a way of reconfiguring human relations with, and in, the material world. This is a concept that focuses not on the rights or limits of bounded entities, but on the lively flows that make life possible. Convivial dignity is experienced as a form of relational togetherness where entangled existence is recognised as being inescapable. This recognition does not necessarily induce desires for attachment nor detachment. It certainly doesn't necessarily prompt nor rely on an ethic of care. It is a recognition of the need to attune and adapt to this togetherness if we are to continue to live together on this planet. These macro-level arguments are brought to life through attention to the micro-level. To ease you into the journey I make between multiple concepts, disciplines and scales, the final section of this introduction outlines how I intend to meander through these concepts to enliven the rather bold claims I am making here. I invite you to join me on this journey and eat from this bountiful menu.

Mapping the flows

While the fieldwork is the heart and soul of this book, I begin by attending to the theoretical foundations and approaches to eating that inform my work. As ever, these claims are not an attempt to discount alternatives, but to build a 'tool box' to support more sustainable futures. Chapter 2 begins with close attention to eating as a 'starter' to prime the appetite to sense the ways interactions with food can disrupt myths of human exceptionalism. With a little sustenance in our minds and bellies, I then connect the focus on food to broader debates about the Anthropocene, the humanist subject it centres on and the necessity of becoming attuned to uncertainty and developing the skills and capacities of adaptability in times of climate change. I contend that dominant notions of human exceptionalism, as embedded within anthropocentric logic, must be rethought and 'redone' to facilitate the becomings of new modes of being and doing responsive to our human and more-than-human entanglements. In line with many in the fields of science-and-technology studies and geography, experimentation is explored as a variable mode of practice capable of generating these alternatives. I suggest the need for a nuanced approach to experimentation, identifying the notion of playful tinkering as a variation in the experimental spirit.

Playful tinkering is presented as involving the need to enter into potentially risky situations out of which—if the 'charge' (Rose, Cooke & van Dooren, 2011) of play has ignited something new—more than the sum of the individual nodes in the relations, springs forth. Play is experimental, but it does not invoke the spectre of the expert. It is open and accessible. As I have suggested, play is also not predicated on the formation of attachments. Attachment can be highly useful and induce significant affective responses that drive action, but I question whether attachment is always necessary in the cultivation of responsiveness and attunement. Indeed, play can also be facilitated through desires for detachment (such as a desire to engage with strange, unpleasant matter; the poking of a dead bird found in the garden riddled with maggots or being consumed by birds of prey, or the turning of a compost heap riddled with rats, their faeces and urine). These encounters draw you in at the same time as they repulse. These moments can also be marked simply by recognition of having to 'live together'. Doing this well, I argue, can unfold when guided by the ethico-political notion of convivial dignity that draws attention to mutual vulnerabilities. As such, playful tinkering enacted with convivial dignity is offered as a way of understanding and contributing to the development of tools—practical and discursive—to amplify relational togetherness in times of uncertainty through particular ways of being and doing. Having mapped out these broad ontological and world-making concepts, I then turn my attention, in Section 1, back to food to begin our practical explorations of new ways of being and doing in the Anthropocene through close attention to taste.

Chapter 3 puts sensorial engagement front-and-centre through discussion of the contradictory natures of taste which I use to exemplify key debates about human/nature relationships in the contemporary era. To do this, it highlights the anthropocentric modes of thinking invoked by linking taste to distinction and compares these to the growing significance attached to the gustatory experience of taste. Here, following those working in the arenas of the visceral, relational and affective turns, I suggest the need for a more generous and generative approach to taste, one which understands it as being induced by multispecies, multisensorial encounters. Indeed, taste is shown to be open and productive, configured through relations among biological response, material attributes of foodstuffs and social, economic and environmental conditions. While taste may become congealed at particular points in time, it is always a practice of becoming-with produced by flows of, among and exceeding bodies, materialities, experiences and temporal-spatial confines.

Taste is shown to be, first and foremost, relational. I argue that particular acts of playful tinkering with food provisioning motivated by embodied, sensorial experiences (rather than solely governed by economic and time constraints) and attunement to the togetherness-in-relation that produces taste, can lead to an openness to being moved by taste, particularly when guided by convivial dignity. This openness and responsiveness to relational flows of life are suggested to be more generative of alternative ways of being and doing in the world than are conceptions of care entrenched in anthropocentric humanist notions that rely on formation of connections and attachments. These tastes for playful tinkering guided by convivial dignity, I contend, can support an ecological awareness that seeks to counteract destructive modes of human mastery that appear to characterise the Anthropocene. This lays the foundations for the subsequent chapters in this section which draw on in-depth fieldwork to identify the affordances that support the development of the skills and capacities needed to attune and respond to this 'togetherness-in-relation' and the potential these have to reconfigure anthropocentric behaviours and possibly the figure of the Anthropos most commonly conceived of by this term. This fieldwork occurs in three key settings: urban food gardens, Alternative Food Networks and at the Royal Canberra Agricultural Show.

Drawing on research walking and talking with people in their food producing gardens, in Chapter 4 I demonstrate the ways in which taste motivates particular food habits and practices. Access to food producing spaces, nostalgia for tastes of their past and exposure to a variety of tastes in varied geographical locations are all shown to encourage the participants' engagement in experimental, tinkering play with food in the present, with an eye to the future. Attunement to the relations through which taste is produced and a responsive openness to being moved by its affective force is shown to position taste as a potentially productive basis for supporting alternative ways of being and doing in the Anthropocene. Indeed, participants are found to engage in practices of making do which are driven by recognition of the potential for the reconfiguration of relations among matter underpinned by the capacity to see 'life left' in things. This, I suggest, can provoke more environmentally sustainable ways of living even when people are not motivated by ecological sensibilities.

Chapter 5 builds on the previous chapter's identification of the affordances that encourage people to 'be moved' by food through playful tinkering to focus on how attunement to taste facilitates responsive relations of togetherness predicated on convivial dignity. It does this through discussion of the way food flows through the homes of those who source their food via Alternative Food Networks (AFNs). Here, I attempt to demonstrate how an ethic of convivial dignity captures the participants' recognition of relations of togetherness that exceed connections among bounded entities. In this way, convivial dignity is shown to move beyond care and concerns with attachment as ways of categorising and motivating environmentalsensibilities and low-resource use. It is the affective force of recognition of togetherness-in-relation—not the matter in and of itself—and the impact of the affective atmospheres within which they are brought together (which, in this chapter, is investigated in relation to the AFNs and geographical notions of place which are shown to stretch beyond the events and locales of where the food provisioning occurs) that is shown to be capable of providing opportunities to develop a taste to attune to non-anthropocentric relations marked by convivial dignity.

The final chapter in this section extends the discussion of the affective force of taste as produced through responsive relations of togetherness linked to place to an unlikely site—Agricultural Shows. Agricultural Shows have historically been understood as being underpinned by discourses of colonialism and modernity marked by socio-technical progress. However, through engagement with small-scale growers competing in the horticulture section of the show and through observation and discussion with the judges, the limited capacity of anthropocentric grammars predicated on notions of human-exceptionalism to capture the relational interactions with taste in these sites is brought to the fore. These limits are experienced through the non-narrativisable, or 'more-than' moments that occur in multisensorial encounters with both the formation of aesthetic and gustatory taste in judging and growing food for exhibition. Here, I suggest we see evidence of how a 'training of sensitivities' in particular practices in particular places can enable the emergence of responsiveness to unexpected tastes formed through recognition of togetherness-in-relation. This training, I suggest, occurs in a playful tinkering approach and is capable of encouraging a reconfiguration of human subjectivities and non-anthropocentric modes of being and doing when enacted with the alternative ethico-political grammar of convivial dignity. That this occurs in such an unexpected site ensconced as Shows, which are commonly understood to be sites that promote representations of agriculture as the bastion of scientific progress, is suggestive of the ways in which attunement to the relational production of taste offers opportunities to destabilise and re-materialise daily habits, behaviours and beliefs in ways that can support and amplify non-anthropocentric ways of being and doing.

As we move into the second section of the book, we shift our attention from the seemingly more pleasurable arena of taste to that of waste. I argue that the murkier realm of food waste and particular forms of multisensorial, visceral engagement with food surplus and/or its avoidance—engagements marked by playful tinkering guided by convivial dignity—can also promote creative, resilient modes of being and doing in the world that exceed the limits of anthropocentric imaginings. In Chapter 7, I highlight the need for us to rethink the productive potential of food waste by mapping out its associated social, environmental and economic issues and the key points where waste occurs in the food system. I then demonstrate that while there has been significant NGO and government attention to these issues in recent years, there is a tendency for these to reproduce anthropocentric norms predicated on human exceptionalism. These are shown to be particularly evident in the invocation of narratives of human-centred care that obfuscate the relational human/more-than-human entangled becomings of waste. I argue that to sustain ecologically sensitive behavioural and attitudinal changes to food waste, the food system and, more broadly, to our ways of living, there is a need for a systematic narrative shift that supports articulation of alternative ways of being and doing in the world that pushes beyond the human exceptionalism ingrained in anthropocentric grammars. These grammars need to be not solely reliant on appeals to care more for the environment or those less fortunate than ourselves. A material-semiotic approach that draws sustenance from the experimental practices already in play in our communities is needed.

As carried out in the Taste section, the potential of alternative ways of being, doing and grammars that give expression to these is explored in the remaining chapters of this section through fieldwork directly related to food waste. Overarchingly, I contend that the practices of the food waste minimising gardeners, shoppers and food rescue workers and volunteers I focus on in the forthcoming chapters involves play enacted in a tinkering approach. I suggest that through these means the participants can become attuned to the relations through which waste is produced and managed. Just as taste is shown to be in-process and relational, here we see that waste never just is, it only becomes.

Via responsiveness to the relational, lively flows or togetherness-in-relation of food waste, mutual vulnerabilities are recognised, the necessity and vitality of multispecies entanglements are experienced, and the life of materials prolonged. This is particularly pronounced in encounters with scarcity and abundance. Through discussion of the modes of being and doing of the participants in this research, I suggest that we can discern sites where people are developing the skills and capacities to adapt to future uncertainty and enacting alternative ontologies capable of supporting more sustainable futures. While these are experienced through relations among materials, the need to amplify these alternatives is material-semiotic in nature, demanding new grammars that enable us to destabilise the exceptionalism of the Anthropos in the Anthropocene. Playful tinkering guided by convivial dignity is one attempt to do just this.

To begin this process, in Chapter 8 I highlight the affective force of waste as experienced by gardeners and AFN participants in their everyday lives. Through discussion of the way food flows through their homes, waste minimisation strategies are shown to be prompted by attunement to multispecies entanglements, recognition of mutual vulnerabilities and attempts to anticipate and respond to the transformations these induce. These attunements are supported by the practices of moving and being moved by food, and through responsiveness and adaptability to the contingencies experienced in times of both abundance and scarcity. While there is some overlap, many of these interactions with food are shown to deviate from the food waste minimisation tactics and strategies promoted in food waste reduction campaigns. The creative, visceral, responsive actions of participants in this research are shown to be developed in a playful tinkering mode that is underpinned by growing openness to relational flows marked by convivial dignity. While these practices are contingent on the presence of particular spatial and social conditions, careful attention to the behaviours of those reducing waste in their homes are shown to be capable of broadening the suite of strategies available to support food waste reduction and encourage less resource usage in the wider community.

In Chapter 9 we explore food waste through encounters with compost heaps, bokashi buckets, worm farms and backyard chooks in the homes of gardeners. While the previous chapter demonstrated the importance of participants becoming attuned to the becomings of food in their kitchens and homes, this tended to manifest itself in a need to carefully manage and limit the transformations of food. For composters, multispecies relations and their transformational potential are actively encouraged eliciting ongoing encounters with shifting forms of life and death and being and doing that exceed bounded, identifiable entities.

Compost is a site rife with forms of togetherness that cannot be understood through simple categories of attachment and detachment and conceptions of care. In these often tense multispecies landscapes, the risk induced by mutual vulnerabilities and the limits of human control are exposed as togetherness-in-relation reigns. These modes of encounter are shown to be generative of forms of being and doing that challenge the human exceptionalism that dominates anthropocentric narratives through the embrace of the uncertainty and unknowability of these sites of non-bounded, entangled becomings-with. The receptivity and attunement encountered here through playful tinkering with food waste is suggestive of a generative form of ethico-political practice in the compost heap. These are practices that are sometimes unsettled by, but always attuned to, life as flows and interdependencies which are experienced as an experiment in alterity-in-relation guided by convivial dignity.

From the microworlds of the compost heap, Chapter 10 scales up our arena of concern to touch on more macro-level practices that occur outside homes. The focus here is on campaigns designed to support a reduction in farm-gate waste by selling ‘ugly’ food in mainstream supermarkets and efforts to redistribute surplus food through food rescue organisations. This shift from the household-level raises questions about the utility of scaling-up responsiveness as a key affordance for encouraging attunement to the multispecies, relational becomings of waste.

Responsiveness at larger scales is shown to provide temporally contingent solutions to surplus that tend to obfuscate the broader systematic problems that produce it. Indeed, not all responsiveness is necessarily generative of non-anthropocentric modes of being and doing. However, as I return to my overarching concern with small-scale practices by attending to the way those who work, volunteer or are clients of food rescue organisations interact with surplus food, I once again highlight the utility of responsiveness to the relational vitality of food in supporting less resource intensive lifestyles through practices of making do.

As we saw in Chapter 4, making do involves engagement in playful tinkering that can support attunement to togetherness-in-relation guided by convivial dignity. It is also capable of promoting development of the skills and capacities to attend to the uncertainty that will be a more regular experience in our food futures. So, while responsiveness in Chapter 10 is identified as a means of large corporations containing their responsibilities to particular temporal periods and deferring their responsibility for the generation of surplus, it is reinforced as a key affordance for encouraging minimal resource usage at the household level. Given the ongoing failures of governments and corporations to act sufficiently to stymie environmental damage, I reiterate the need to focus on small-scale, local actions as sites where we encounter creative and agile ways of living that respond to uncertainty and offer alternative ontologies and world-making practices.

These alternative ways of being and doing are material-semiotic in nature. The text concludes by consolidating the argument developed throughout that identifies the need for new grammars to represent and amplify these productive possibilities. Environmental narratives, notions of care and desires for intimate attachment are unlikely to support the necessary reconfiguration of the Anthropos that will provoke more generative ways of living together. Convivial dignity is offered as an ethico-political guide for encouraging recognition of the inescapability of our togetherness-in-relation. It is concerned with supporting multispecies and interdependent relational becomings-with that give life to alternative ways of being and doing. It is developed throughout the text in a playful, tinkering mode that invites response and attempts to support responsiveness and response-ability to what it means to live in a time of uncertainty.

References

Abrahamsson, S., & Bertoni, F. (2014). Compost politics: Experimenting with togetherness in vermicomposting. *Environmental Humanities*, *4*, 125–148.

Abrahamsson, S., Bertoni, F., Mol, A., & Ibáñez Martín, R. (2015). Living with omega-3: New materialism and enduring concerns. *Environment and Planning D: Society and Space*, *33*(1), 4–19.

AGHS. (2006). *A gardener's city: Canberra's garden heritage*. Melbourne: Australian Garden History Society.

Anderson, D. (2014). *Endurance: Australian stories of drought*. Collingwood, VIC: CSIRO Publishing.

Barthel, S., & Isendahl, C. (2013). Urban gardens, agriculture, and water management: Sources of resilience for long-term food security in cities. *Ecological Economics*, *86*, 224–234.

Bennett, J. (2010). *Vibrant matter: A political ecology of things*. Durham, NC: Duke University Press.

Blecha, J., & Leitner, H. (2014). Reimagining the food system, the economy, and urban life: New urban chicken-keepers in US cities. *Urban Geography*, *35*(1), 86–108.

Carolan, M. (2016). Adventurous food futures: Knowing about alternatives is not enough, we need to feel them. *Agriculture and Human Values*, *33*(1), 141–152.

Gibson-Graham, J. K. (2008). Diverse economies: Performative practices for other worlds. *Progress in Human Geography*, *32*(5), 613–632.

Gibson-Graham, J. K. (2011). A feminist project of belonging for the Anthropocene. *Gender, Place and Culture*, *18*(1), 1–21.

Gibson-Graham, J. K. (2014). Being the revolution, or, how to live in a "more-than-capitalist" world threatened with extinction. *Rethinking Marxism*, *26*(1), 76–94.

Gibson-Graham, J. K. (2015). Ethical economic and ecological engagements in real (ity) time: Experiments with living differently in the Anthropocene. *Conjunctions. Transdisciplinary Journal of Cultural Participation*, *2*(1), 44–71.

Goodman, M. K. (2016). Food geographies I: Relational foodscapes and the busy-ness of being more-than-food. *Progress in Human Geography*, *40*(2), 257–266.

Haraway, D. (2008). *When species meet*. Minneapolis & London: University of Minnesota Press.

Harbers, H., Mol, A., & Stollmeyer, A. (2002). Food matters: Arguments for an ethnography of daily care. *Theory, Culture & Society*, *19*(5–6), 207–226.

Head, L. (2016). *Hope and Grief in the Anthropocene: Re-conceptualising human–nature relations*. Oxon & New York: Routledge.

Head, L., Atchison, J., & Phillips, C. (2015). The distinctive capacities of plants: Re-thinking difference via invasive species. *Transactions of the Institute of British Geographers*, *40*(3), 399–413.

Heuts, F., & Mol, A. (2013). What is a good tomato? A case of valuing in practice. *Valuation Studies*, *1*(2), 125–146.

Hinchliffe, S., & Whatmore, S. (2006). Living Cities: Towards a politics of conviviality. *Science as Culture*, *15*(2), 123–138.

Hird, M. J. (2010). Meeting with the microcosmos. *Environment and Planning D: Society and Space*, *28*(1), 36–39.

Hird, M. J. (2014).

Jarosz, L. (2008). The city in the country: Growing alternative food networks in Metropolitan areas. *Journal of Rural Studies*, *24*(3), 231–244.

Latour, B. (2004). How to talk about the body? The normative dimension of science studies. *Body & society*, *10*(2–3), 205–229.

Law, J. (2009). Actor network theory and material semiotics. In B. S. Turner (Ed.), *The new Blackwell companion to social theory* (pp. 141–158). Oxford: Wiley-Blackwell.

Law, J., & Mol, A. (2002). Complexities: An introduction. In J. Law & A. Mol (Eds.), *Complexities: Social studies of knowledge practices* (pp. 1–22). Durham, NC: Duke University Press.

Law, J., & Mol, A. (2008). *Globalisation in practice: On the politics of boiling pigswill.* Geoforum, *39*(1), pp. 133–143.

Law, J., & Mol, A. (2008). The actor-enacted: Cumbrian sheep in 2001. In C. Knappett & L. Malafouris (Eds.), *Material agency: Towards a non-anthropocentric approach* (pp. 57–78). Dusseldorf: Springer.

Mol, A. (2008). I eat an apple. On theorizing subjectivities. *Subjectivity*, *22*(1), 28–37.

Mol, A. (2010). Actor-network theory: Sensitive terms and enduring tensions. *Kölner Zeitschrift für Soziologie und Sozialpsychologie Sonderhefte*, *50*(1), 253–269.

Mol, A., & Law, J. (2004). Embodied action, enacted bodies: The example of hypoglycaemia. *Body & Society*, *10*(2–3), 43–62.

Mol, A., Moser, I., & Pols, J. (2010). Care: Putting practice into theory. In A. Mol, I. Moser, & J. Pols (Eds.), *Care in practice: On tinkering in clinics, homes and farms* (pp. 7–26). Bielefeld, Germany: Transcript Verlag.

Paulson, W. (2001). For a cosmopolitical philology: Lessons from science studies. *SubStance #96*, *30*(3), 101–119.

Pincetl, S. (2007). The political ecology of green spaces in the city and linkages to the countryside. *Local Environment*, *12*(2), 87–92.

Pitt, H. (2015). On showing and being shown plants-a guide to methods for more-than-human geography. *Area*, *47*(1), 48–55.

Plumwood, V. (2001). Nature as agency and the prospects for a progressive naturalism. *Capitalism Nature Socialism*, *12*(4), 3–32.

Pothukuchi, K., & Kaufman, J. L. (1999). Placing the food system on the urban agenda: The role of municipal institutions in food systems planning. *Agriculture and Human Values*, *16*, 213–224.

Rose, D. B., Cooke, S., & van Dooren, T. (2011). Ravens at play. *Cultural Studies Review*, *17*(2), 326–343.

Sonnino, R. (2009). Feeding the city: Towards a new research and planning agenda. *International Planning Studies*, *14*(4), 425–435.

Steel, C. (2009). *Hungry city: How food shapes our lives*. London: Vintage.

Steffen, W., Crutzen, P. J., & McNeill, J. R. (2007). The Anthropocene: Are humans now overwhelming the great forces of nature. *AMBIO: A Journal of the Human Environment*, *36*(8), 614–621.

Steffen, W., Grinevald, J., Crutzen, P., & McNeill, J. R. (2011). The Anthropocene: Conceptual and historical perspectives. *Philosophical Transactions of the Royal Society of London A: Mathematical, Physical and Engineering Sciences*, *369*(1938), 842–867.

Steffen, W., Persson, Å., Deutsch, L., Zalasiewicz, J., Williams, M., Richardson, K., … Svedin, U. (2011). The Anthropocene: From global change to planetary stewardship. *AMBIO*, *40*(7), 739–761.

Taylor, K. (2006). *Canberra: City in the landscape*. Canberra: Halstead Press.

Tsing, A. (2012). Unruly edges: Mushrooms as companion species for Donna Haraway. *Environmental Humanities*, *1*(1), 141–154.

Turner, B., & Hope, C. (2015). Staging the local: Rethinking scale in farmers' markets. *Australian Geographer*, *46*(2), 147–163.

Turner, B., Pearson, D., & Dyball, R. (2012). *Food in the ACT: A preliminary study of issues for the ACT.* Canberra: Government Environment *and* Sustainable Development Directorate.

Turner, B., Pearson, D., & Dyball, R. (2014). Planning for regional food security: A case-study of the Australian Capital Territory (ACT). *Locale: The Australasian-Pacific Journal of Regional Food Studies*, (4), 20–46.

Whatmore, S. J. (2006). Materialist returns: Practising cultural geography in and for a more-than-human world. *Cultural Geographies*, *13*(4), 600–609.

Zournazi, M., & Stengers, I. (2002). A 'Cosmo-politics'—Risk, hope, change—With Isabelle Stengers. In Zournazi, M, *Hope: New philosophies for change* (pp. 244–273). Annandale, NSW: Pluto Press Australia.

2 An appetiser

Eating, being and playing with convivial dignity

Introduction

This chapter lays the foundations for my contention that in the fieldwork food stories that unfold in later chapters, we can sense new ontologies that offer alternative ways of being and doing in the world. These non-normative modes of being reconfigure the human subject at the heart of the Anthropocene and redirect attention to the relational flows that make life, and new material-semiotic approaches to the world, possible. More broadly, this chapter grounds my suggestion that everyday engagement with food provides a raft of encounters that open up opportunities to cultivate the skills, capacities and agility necessary to adapt to our uncertain futures and work towards mitigating and adapting to some of the effects of climate change. The basic necessity that is food, and the lives and livelihoods of the humans and more-than-humans bound up in its production, distribution, ingestion and waste, makes it ripe for analysis. The fundamental physiological need that means bodies require us to eat provides a starting point to explore our meetings with food. Beginning here enables me to wander through the recent attention paid to the 'eating body' which has laid the foundations for the work exploring the embodied, visceral and affective force of food (see Hayes-Conroy & Hayes-Conroy, 2013; Mol, 2009; Probyn, 2012) that I am to build on.

While throughout this chapter my narrative strays a little away from food at times, as I attempt to introduce and unpack the modes of thinking that I draw on to motivate the ontological arguments developed, rest assured that food remains firmly in my sights and it will not be obscured for long. You will see in later chapters that the empirical food-focused fieldwork fuels and crystallises the world-making concepts introduced here. However, these deviations from food also serve to highlight the relational entanglements of our food encounters—food is not always uniquely identifiable or distinguishable from humans and other nonhumans, nor is food only an actant in and of itself. I am concerned with food's becomings, as forms of matter-in-relation with bodies and other more-than-humans, and the effects of these becomings on bodies and the 'world' more broadly. I wish to focus attention on the relational flows of life and how our multisensorial engagements with food—from taste to waste—can highlight the

DOI: 10.4324/9780429424502-2

relations of togetherness that produce our worlds and how particular forms of attention to these relations can support and amplify more responsive forms of living that disrupt conceptions of the modern subject. I initiate this by focusing on the eating and metabolic body and the consequent encounters with the fragility of anthropocentric notions of human exceptionalism we find in the multispecies and multi-material process of digestion. Recognition of this fragility prompts the need for alternative ways of being with the world that are responsive, attuned and, as we shall see, also playful.

In the second half of this chapter, I outline the notion of playful tinkering as a variant of experimentation that I suggest enlivens the non-anthropocentric ways of being and doing that we will encounter in the fieldwork that populates the following chapters. Engagement in play is shown to be capable of enacting a form of 'ontological choreography' (Haraway, 2008b) that, through its responsive nature, attends to and produces relations of togetherness or what I refer to as togetherness-in-relation. Production of this choreography is not limited to bounded entities or knowable matter. Play exceeds these limits, as the focus is on the relations that enable play. As Haraway writes, 'play involves efforts to determine what is this, what can it be, how can we be together?' (Haraway, 2008b). Play facilitates practices of world-making where the human subject can be reconfigured and behaviours shifted.

However, the generative nature of play may not be sufficient to alter our current trajectory where we find ourselves hurtling towards further environmental destruction. It is also not the only characteristic of the interactions that I discern in the gardens, kitchens, exhibition halls and compost heaps of those involved in this research. These are not only material practices but discursive as well. In our efforts to promote less resource intensive modes of living, particularly around food, much of the battle also needs to occur at the narrative level. We not only need non-anthropocentric ways of being and doing in the world but we need narratives capable of representing and amplifying these. Convivial dignity is the nomenclature and concept I develop here. I see those participating in this research as developing these alternative modes of being through responsive modes of playful tinkering that acknowledge the necessity of their togetherness with human and more-than-human others in the world and are attuned to the relations that configure these. These are not necessarily relations that are marked by attachments nor detachment. Nor are they reliant on an ethics of care. Instead, they are much more generative, invoking multifarious, open forms of togetherness predicated on understanding and living the untruths of human exceptionalism and the necessity of our relational existence. Convivial dignity is offered as an alternative grammar for representing, and working towards realisation of, these new ontologies.

Digesting the material world

If we are to promote new ontologies that support less resource-intensive lifestyles, then food offers us a uniquely intimate and familiar place to start. Our engagements with food—from taste to waste—are sites where relational entanglements among humans and nonhumans are often quite conspicuous. Eating, gardening and composting, to name a few of the food related activities we will explore in this text, all rely on overt engagement with the material world. They are practices that generate transformation of humans and more-than-humans alike through relational interactions. Yet our tendency in the minority world is to understand these as practices within which the human actor is central. Human choices, decisions and labour are typically viewed as the key elements enabling and controlling these transformations. Of course, close attention reveals a raft of more nuanced and complex stories at play. At the risk of over-simplification, at the heart of many of these is the fact that human bodies need food and food induces reactions from, in and with these bodies. These reactions can range from affective responses (such as the desire for certain tastes or disgust at the mould growing on leftovers forgotten at the back of the fridge) through to physical form (such as accumulation of fat, sheen of hair and strength of nails and, of course, maintenance of life). Throughout this text, I will argue that it is this very embodied and visceral engagement with food, and it's capacity to support our attunement to the relational ways in which these are produced, that can support alternative, non-anthropocentric ways of being and doing in the world. Here we jump right into the guts of the matter with an appetiser made up of the physiological and biological processes of eating.

While eating cannot be removed from social, political, cultural and spatial concerns (we know bodies can become attuned to different tastes in different contexts), they are processes over which humans largely lack control. We cannot manipulate our digestive systems (Mol, 2008) by attempting to train the various gastrointestinal components to perform in a particular way (perhaps, except in extreme examples such as sword swallowing or with external assistance such as bacteriotherapy). Kuriyama points out that this inability to be 'trained' positions the digestive system as distinct from externally viewed muscular components of the body. While the muscular body has come to represent the will, control and autonomy of the individual Cartesian subject, the metabolic body belies the great myth of human supremacy and mastery over the 'natural' world (Kuriyama in Mol, 2008, p. 29). As Mol writes '[a] person cannot train the internal lining of her bowels in a way that begins to resemble the training of her muscles' (Mol, 2008, p. 30). Thus, in this familiar everyday realm of eating and digesting, the inadequacy of the dominant ways humanism configures human subjects in the world (Hinchliffe & Whatmore, 2006) is brought to the fore. To be human is to be reliant on more-than-human entities. Without them and the relations through which human and nonhuman components interact, there is no way of nourishing human bodies. And just how these relations unfold and what they enact is beyond our control.

Blurring the boundaries of the human: attending to the 'metabolic body'

Eating exposes the limits of human mastery. As Derrida states, 'one never eats entirely on one's own' (1991, p. 115). As we chew and swallow, food shifts in form and shape in response to a multiplicity of responsive actors and metabolic rhythms encountered on its journey. Close attention to eating unsettles the perceptions of a clear divide between humans and more-than-humans. There is a flow of entities encountered in digestion, as Bennett notes, 'all bodies are shown to be but temporary congealments of a materiality that is a process of coming, is hustle and flow punctuated by sedimentation and substance' (Bennett, 2010, p. 49). Food masticated in the mouth travels through the oesophagus into the stomach, then the small intestine where bile and numerous digestive enzymes are released and nutrients extracted for transfer throughout the body, while waste products move through to the large intestine. These multifaceted assemblage processes do not occur in the same way in all bodies, nor in the same way to all foods in each unique body. The form and quantity of the food we eat and fluctuations in our gut flora are some of the components that influence the journey from A to B. Just what will happen is never fully predictable. Food allergies, for example, may incite a body to rally its defences initiating a series of often unexpected actions and reactions. Strong affective responses—'food loathing' (Kristeva, 1982)—can similarly manifest in physical reactions. Kristeva's multi-layered emotional, intellectual and physical response to her lips coming into contact with skin on warm milk draws attention to this. Through eating, and choices not to eat, we can see that food and drink '... taste good or bad, have a nice or gruesome texture. They are, not as delegates of people, but all by themselves, objects of longing or aversion' (Harbers, Mol & Stollmeyer, 2002, p. 217). Food does things. It calls bodies into action, physically and emotionally. The food we choose to take into our body continues to 'do things' through the mechanism of digestion by engaging multiple actors in a series of relational actions and reactions beyond our control, the processes and results of which may well surprise us.

As eaters, we are unable to sustain our selves as whole subjects, distant and different from more-than-human entities. We willingly invite that which is not us to enter into our bodies. Some then become us, or perhaps more accurately, we become with each other. These are fluid processes that cause the various participants to shift in shape and form, congealing, extracting and separating in ways beyond our control and outside of the agency of any one specific actor. Digestion is enacted by a multiplicity of entities engaging in intricate, shifting and responsive relationships. In fact, as Mol and her co-researchers highlight, the search for a dominant actor risks obfuscating the interactions among the various players and how the very form of these entanglements make possible the performative actions of the entities involved '[a]cting and being enacted go together' (Law & Mol, 2008, p. 58).

It is this mutual enactment that Mol and her co-researchers (Harbers et al., 2002) argue is missing in versions of new materialism (focusing specifically on Jane Bennett's work and her identification of the 'thing-power' of omega 3 which is discussed further in Chapter 3). They claim that new materialism's rush to ascribe agency to more-than-humans to destabilise anthropocentric-thinking runs the risk of presenting a far too simplistic account of just how material comes to matter. The relational aspects can become side-lined when one's focus is fixated on identification of an individual actor, whether this be human or nonhuman. As such, Abrahamsson, Bertoni, Mol and Ibáñez Martín (2015, p. 6) suggest the need to move from conceptualisations of action that focus on specific actors and assumptions of agency to explorations of 'modes of doing'. In this way, action becomes 'creatively retheorised'—and able to be conceptualised in alternative ways 'such as affording, responding, caring, tinkering and eating' (Abrahamsson et al., 2015, p. 3). However, the enduring legacy of humanism's modern subject makes such shifts particularly challenging.

The discursive rendering of the 'I' that marks the modern subject of the Anthropocene is not at home in the multispecies account of human bodies that we have thus far encountered in eating and digestion. The metabolic subject cannot willfully act by itself (except, of course, by choosing not to eat or to eat foods with inadequate nutrition, making death—though there are many ways death can unfold—inevitable). 'How to give words,' Mol asks, 'to this mode of being a subject?' (2008, p. 30). Such words, narratives and discourses are difficult to find because such modes of being challenge dominant representations of precisely what it is to be human. Indeed, we need alternative ways to conceive of being and doing in the Anthropocene.

Alternative subjects in the Anthropocene: narrativising material matters

The notion of the modern human subject, or what Plumwood refers to as the Western 'narrative self' (1996, p. 2), sits at the heart of the Anthropocene and fuels discourses of hyper-separation between humans and non-humans. While debate continues about whether or not to accept the notion of the Anthropocene as an official geological era, the meaning and utility of the term also remains subject to ongoing discussion. Dibley (2012) nominates seven versions of the Anthropocene revolving around two key issues, firstly the so-called scientific debate centred on questions of whether there is enough evidence to have the term officially accepted and, if so, when the era began, and secondly its social and cultural framings. The latter are, perhaps, most often initiated not by climate change sceptics but by academics deeply engaged in more-than-human worlds and attuned to the devastating impacts of human-induced climate change (Buck, 2013; Ginn, 2015; Head, 2014; Robbins & Moore, 2013). Head raises two further potential risks. Firstly, she warns against perpetuating a perception of the Anthropocene as a linear and teleological era where contingencies and rupture in geological timescales are glossed over. Such a perspective threatens to position the Anthropocene as just another era, potentially reducing its potential to be generative of alternative modes of being and doing. Secondly, she observes that the 'anthropos at the core is a slippery concept' (Head, 2014, p. 114), pointing to the enduring spectre of human exceptionalism that permeates recognition of a human-induced problem which both escapes human efforts to contain and control it (Head, 2016) while continuing to position 'modern subjects' as having the capacity to act, seemingly unilaterally, to stymie our impact.

Dominant conceptions of the Anthropos are also problematic due to their tendency to flatten out the species scale. Not all of us have violently imposed our will on the planet; in fact, many humans have enacted vastly different ways of engaging with the world, not least around the ways in which they provision food (see, in particular, work carried out on aboriginal Australians from Anderson, 2007; Bawaka et al., 2015; Rose, 1999; Weir, 2009). However, as Ginn writes, 'the very act of asking the question, "Is this the Anthropocene?" demonstrates that we have moved into an era of anxiety about the prospects of planetary life' (Ginn, 2015, p. 352). In my discussion of our relationships with food, I begin from the position that there is overwhelming evidence that human-induced climate change is real and that, without alterations to the way we live and to our very enacting of humanism, the future is bleak. I want to tread carefully here by emphasising that humanism is not simply a product of Cartesian metaphysics. It has also been fuelled through recourse to materiality (Anderson, 2014). As such, the material turn in and of itself does not challenge human exceptionalism. Indeed, narratives of human exceptionalism, as Anderson shows, have been fuelled by distinctly material manifestations as evident in the work of 19th century comparative anatomists and their attention to the body in what she calls 'anatomical humanism' as well as in more contemporary geo-engineering projects.

If materiality is already embedded within humanism, then to shift anthropocentric thinking we need to not only draw attention to matter but attend to it anew. We must not obscure the materially informed contributions to the genesis and ongoing stranglehold of humanism over thought and practice, but explore how encounters with material-semiotic practices can be used to propose different forms of humanism. As Plumwood writes:

> If our species does not survive the ecological crisis, it will probably be due to our failure to imagine and work out new ways to live with the earth, to rework ourselves and our high energy, high consumption, and hyper-instrumental societies adaptively...We will go onwards in a different mode of humanity, or not at all.
>
> (Plumwood, 2007, p. 1)

Transforming ways of being and doing in the Anthropocene: the affordance of making do

The potential for creative reconfigurations of our 'mode of humanity' is already manifest in numerous small-scale endeavours around the globe, including the majority world, where people actively demonstrate their capacity to live and conceive of themselves outside of the modern, 'Western narrative self' particularly in relation to how we eat, drink and protect ourselves from the elements (Atchison & Head, 2013; Gibson-Graham, 2008, 2011; Head, 2016; Head, Atchison, &

Phillips, 2015). While Head's work identifies that many of these endeavours are primarily enacted in times of scarcity, abundance or disaster, the witnessing and documenting of these actions demonstrates that humans do indeed have multiple skills and capacities to respond and adapt to disruption and uncertainty enlivened and amplified by particular affordances. Many of us act differently when we must 'make do' in response to our visceral, embodied encounters with recalcitrant matter and the generative potential of the relations that unfold.

Studies into human engagement with water, both its lack in times of drought and its abundance when floods devastate and displace, are riddled with documentation of these capacities (see Allon & Sofoulis, 2006; Head, 2008; Whatmore, 2013; Whatmore & Landström, 2011). In this work, we see evidence of people taking action in their homes and connecting with their communities to develop more relational modes of living attuned to their more-than-human and human counterparts. This prompts action ranging from collecting shower water in buckets to participating in collaborative flood research. The global food sovereignty and food justice movements also involve expansive community-level, coordinated initiatives enacted in many forms ranging from urban foraging, establishment of community gardens and development of localised food systems. The diversity and breadth of these small-scale, local actions contrasts with the significant lack of action we see in local, national and global governance on climate change. While we should most certainly not give up on the importance, symbolically and practically, of policy-led approaches, a focus on small-scale capacities to engage and do things differently should be emphasised in the face of the limited and stalled response of those in political power. Here I concur with Head when she writes:

> At a time when top-down intergovernmental action seems not to be up to the task, survival may depend on more localised vernacular understandings and practices. Important intellectual resources come from places understood as marginal to environmental preservation; indigenous engagements, gardens, suburbs, farms, domestic homes. We can revisit empirical evidence from these to consider capacity and vulnerability in new ways.
>
> (Head, 2016, p. 13)

Small-scale actions—particularly when we are attending to food and the inherent problems of international agri-business—are often ridiculed and dismissed as being politically and practically insignificant. The diverse economies work of J. K. Gibson-Graham that attempts to map and document initiatives people engage in outside of mainstream capitalist logic has borne the brunt of much of this criticism. However, as Gibson-Graham points out, engagement in these diverse economies 'account[s] for more hours worked and/or more value produced, than the capitalist sector' (2008, p. 617). For some, such evidence encourages conceptualisation of the contemporary geological era as the capitalocene (Haraway, 2015; Malm & Hornborg, 2014; Moore, 2017) positioning capitalism, rather than the Anthropos, as the dominant actor inducing climate change. For Head, (2016), even less time and value will be derived from the capitalist sector in the future when people will be forced to invest more of their labour (and time) to accessing increasingly limited resources such as food, water and energy on a daily basis.

The global presence of numerous forms of diverse-economies (Gibson-Graham, 2008) supporting extra-capitalist ways of 'doing' and 'being' in the world and evidence of peoples' capacities to 'manage' and reconfigure their daily life practices in times of scarcity and abundance (Head, 2016) indicate that ontological projects subverting the supposed norms of the Western modern subject are already under construction. They remain, however, marginal and small-scale. Yet, as concerns grow about the predicted dystopic future wrought by climate change, anxieties about how we will be able to live—starting with the fundamental concern of how we will feed ourselves—come to the fore. It may be that, as Head writes, 'gardeners and practitioners will be our instructors' (2016, p. 128) on how to live differently in the Anthropocene. These instructors will not be prescriptive but will be willing to share their 'learnings'. The challenge now seems to be how such practices can be supported and amplified. What will work cannot be known from the outset. It requires engagement in multiple forms of experimentation with both matter and narrative. In the following sections, I provide an entrée into the ontological arguments that underpin the experimentations explored and enacted throughout this book. While food is not the focus here, my aim is to develop a way of conceptualising the particular forms of action and interactions among humans and more-than-humans that I identify in the food-specific encounters of my fieldwork and to highlight how these contribute to the development of modes of being and doing that challenge anthropocentric practices and narratives.

Material-semiotic experimentation: the pursuit of playful variations

Experimentation is critical for living in the present with an embodied and engaged attunement to our impacts on the future. The specific form and features of this future may well hinge on the capacity for us to be and do differently in the world. As Gibson-Graham writes:

> The spatiality of production, consumption, reproduction and exchange is potentially up for grabs as new technologies, old ideas and the vital capacities of human and nonhuman agents combine in experiments with living in a different mode.
>
> (2011, p. 12)

Such experiments cannot simply be enacted through human-centred processes. They may involve looking to the past to relearn practices necessary for the future (Head, 2016). They will certainly involve close attention to present practices to bring into view and extend existing skills and capacities while also identifying vulnerabilities and how we can work these differently (Gibson-Graham, 2008, 2011; Head, 2016). These experimentations that imbricate the future in their present practices must be guided by a sensitivity and responsiveness to the world. In this way, experimentation needs to be accompanied by Latour's conception of 'learn[ing] to be affected, meaning "effectuated", moved, put into motion by other entities, humans or nonhumans' (Latour, 2004, p. 205)' and Haraway's notion of 'response-ability' (2008b) being:

> . . .the cultivation of the capacity to response in the context of living and dying in worlds for which one is for, with others. So, I think of response-ability as irreducibly collective and to-be-made. In some really deep ways, that which is not yet, but may yet be. It is a kind of luring, desiring, making-with'.
>
> (Haraway, 2014, pp. 256–257)

The task then, is one of working out how to experiment sensitively with others as a means of becoming attuned to how we can be, become with, make-with and belong in the world in more sustainable ways.

Before we go too much further along these lines, let me acknowledge that my use of variants of the term 'experiment' run the risk of committing the 'lexical laxity' Lorimer and Driessen (2014, p. 178) caution against when commenting on the term's usage in geography in recent years. Indeed, they note, 'it is sometimes difficult to conceive of what social changes are not experimental in this literature' (2014, p. 178). I draw on experimentation to connect my ideas to the body of research in a number of fields, most notably human geography and science-and-technology studies. Stengers, (1997) work on experimentation is key here, but the concept I develop is less explicitly focused on scientific practice in the lab or the field. While I am interested in practices that 'force thought', I also pursue what thoughts 'force practice'. That is, how can particular narratives or grammars represent and amplify specific ways of doing and being. While I take experimentation as a starting point, aiming to build on Whatmore's calls for posthumanist thinking to take 'a more experimental tack', (2006, p. 34) my destination is the identification of playful tinkering as a variant of experimentation that I argue is generative of new ways of being and doing. This is particularly apparent when play is enacted with the ethico-political notion of convivial dignity. These concepts, which I develop throughout the remainder of this chapter, attempt to attend to material-semiotic ways of supporting more agile, less resource intensive modes of living in times of uncertainty.

Tinkering with play: the limits of trust, attachment and care

Play is a familiar term and form of engagement that, while not uniform, simple or singular in meaning, can be considered to be part of human and animal (notably mammals) repertoires of skills, capacities and affordances. The familiarity of play, the lack of presence of the 'spectre' of the 'expert' and its capacity for world-making through responsive human/more-than-human interactions marks play out as a more accessible, translatable and, in particular contexts, more generative variant of traditional notions of scientific experimentation. The development of playful tinkering here is not an attempt to displace or supersede the work of experimentation, instead it builds on Cameron's calls for 'more open, even playful forms of experimentation to try out new ways of living in the Anthropocene' (Cameron, 2015, p. 100). To this end, this notion of playful tinkering represents an effort to capture, convey and enact the openness, drive and creativity that productive forms of being and doing in the Anthropocene require. Like broader notions of experimentation, playful tinkering lacks a specified goal and it is unclear what these 'new ways of living' will look like. Unlike experimentation, however, play does not lend itself to reproducibility so these 'new modes of living' are constantly having to be reworked and developed through responsive entanglements. These, as we will see, tend to unfold in a tinkering fashion that manifests in incremental, 'bit by bit' adjusting (Mol, Moser & Pols, 2010, p. 14).

Tinkering can be understood as a form of experimentation in how to be and do in the world. It is staunchly anti-teleological and without strategy (though not necessarily without purpose). As Mol writes, tinkering 'suggests persistent activity done bit by bit, one step after another without an overall plan. Cathedrals have been built in a tinkering mode, and signalers or aircraft designers also work in this way' (2010a, p. 265). Such processes inevitably involve failures, but it is only through doing that these can be encountered and alternative possibilities cultivated. This does not simply happen by chance, but is instead a result of processes of attunement, versions of learning to be affected and response-ability developed through ongoing interactions among humans and more-than-humans where limits of materiality are regularly encountered. 'Bit by bit' experimentation cannot subvert material limits. Such encounters with matter pushing back confirm the necessity of the ongoing ontological project of learning how to be and do in ways that are responsive to our relational living. It is about how to live together.

Tinkering can create grand structures, impact on the safety of our global transportation networks and produce new foods. Yet, representing these practices and engagement with the material is particularly challenging. In fact, Law and Mol state that '[d]iscursive justifications always betray the specificities of tinkering' because tinkering 'has less to do with thought. But is more a matter of matter. Of the body, of practice' (Law & Mol, 2002, p. 101). While tinkering offers us glimpses into how it is possible to be and do with the world in necessarily relational entanglements attuned to the material, these terms also highlight the very limits of our discursive representations. The spectre of the modern subject remains. As Mol herself notes, the term 'tinkering' and others she calls up such as 'doctoring' and 'caring' are to be used cautiously as they 'suggest that there is a tinkerer. . .separate from the object/subject being tinkered with. . .' (Mol, 2010a, p. 265).

While such a subject has been significantly destabilised in numerous theoretical realms from Marxism to feminism, its persistence is evident in the discursive politics of climate change and the concept of the Anthropocene itself—by deniers as well as in some forms of environmentalism and/or conservationalism. The notion of conscious human mastery and control endures. However, as Heuts and Mol (2013, p. 139) point out, tinkering 'is not a matter of taking control', rather it actually demonstrates the very limits of human control as well as those of the materiality of the more-than-human elements involved (Heuts & Mol, 2013; Mol,

2010a, 2010b). This form of doing, then, challenges dominant conceptions of a 'modern' Cartesian subject consciously acting on the world in the pursuit of progress (the tinkerer here cannot do what he/she likes. Matter pushes back). However, human and nonhuman limits, 'can only be experimentally discovered in the process of tinkering' (Heuts & Mol, 2013, p. 138).

Like tinkering, play also requires action. It is a doing that is only enlivened when players (human and nonhuman) enter into playful encounters. It is the assemblage relationships of play—the actors, conditions and resources—and responsive attunement to other entities that enables it to unfold. While Jane Bennett's notion of naïve realism focuses on the need to mentally cultivate human openness to thing-power in order to promote reconfigurations of human exceptionalism, from the outset, play is more embodied, entangled and active. Play cannot come about simply due to a participant's openness to it, though this is needed. The focus then in play is on relations—entangled collectives—and the generative capacity of the shifting interactions among these. These are features, too, of the string figure game cat's cradle that Haraway offers up as a metaphor for knowledge making and becoming-with in the world. As she writes:

> Cat's cradle is about patterns and knots; the game takes great skill and can result in some serious surprises. One person can build up a large repertoire of string figures on a single pair of hands; but the cat's cradle figures can be passed back and forth on the hands of several players, who add new moves in the building of complex patterns, cat's cradle invites a sense of collective work, of one person not being able to make all the patterns alone. . .If we do not learn to play cat's cradle well, we can just make a tangled mess.
>
> (Haraway, 1997, p. 268)

The potential to end in a tangled mess is indicative of the responsive, convivial, physical and mental work required in this game.

Taking cat's cradle as a form of playful tinkering, we see that while it can be performed alone it is enlivened by exchanges with others, the repertoire of action expands as many hands work with string while, simultaneously, growing numbers of hands and formations increase the risk of a 'tangled mess'. Like play, more broadly, the forms made by the knots and turns in cat's cradle hold only while embodied and enacted when in play. It is in the moment of playing, and particularly through engaging in the risk of becoming-with by encountering limits and relational possibilities enabled through playful tinkering, that possibilities take shape. Indeed, as Haraway writes, play is 'one of those activities through which critters make with each other that which didn't exist before, it's never merely functional; it is propositional. Play makes possible futures out of joyful but dangerous presents' (Haraway, 2014, p. 260). It is, as Massumi writes, 'inventive', 'a veritable laboratory of forms of live action' (2014, p. 12).

Playful tinkering, then, is generative precisely due to the element of risk involved. This begins simply with the risk of not knowing what will happen, of relinquishing control. For Haraway, risk becomes mitigated over time, through repetition, in which players begin to trust that those they engage with will abide by the rules of the game. When they don't, trust is violated and they are no longer 'in play' (Haraway, 2008b). This building up of trust seems to imply that one can come to 'know' those they play with (Haraway writes extensively of her relationships with her dogs through training and practice). However, Rose, Cooke and van Dooren observe that in their participation 'in play' with ravens in Death Valley, trust and assumptions of knowingness or familiarity were not significant aspects:

> In place of the 'open' in which two or more creatures 'get it' together [a la Haraway and her show dog], this raven play was pervaded by a deep sense of unknowing, of distance and mystery. Something quite profound, we thought, occurred in that space of unknowing.
>
> (Rose, Cooke, & Van Dooren, 2011, p. 335)

The capacity to engage responsively—or playfully—with unknowable entities, what Tim Morton calls 'strange strangers' (2010), may be a critical skill we need to cultivate and support to challenge anthropocentric thinking through new modes of being (becoming-with) and doing. Such alternative beings and doings do not necessarily require empathy, attachment or care. Indeed, such affects and practices can be difficult to cultivate with unknowable otherness. Engagement with such extreme alterity presents an intensified risk of the unknown. Ugly, slimy, smelly and 'destructive' creatures can be hard-to-love. As Gibson-Graham observe:

> While we might feel love for other earth creatures and want to accept a responsibility to care for them, might we also extend our love to parasites, or inorganic matter, or to the unpredictability of technical innovation?.
>
> (2011, p. 7)

Not only are attachments, love and care challenging to cultivate with some more-than-humans (I would suggest that the microorganisms that enliven soil to assist food production, the parasites in our guts that aid extractions of nutrients and the multifarious entities involved in food waste decomposition fall into this category), they may also not be the most productive modes of relations. As Bennett suggests, a predilection for these forms of connection can simply reinforce anthropocentric norms where humans must undertake the task of caring for, or being the stewards, of the earth and its creatures. Such modes of relations can also obscure alterity for, as Derrida notes, 'once you grant some privilege to gathering and not to dissociating, then you leave no room for the other, for the radical otherness of the other, for the radical singularity of the other' (Derrida & Caputo, 1997, p. 14). As such, we need to look beyond attachment in our efforts to support more ethical human/more-than-human relations attuned and responsive to uncertain futures. Detachment and other more multifarious forms of togetherness may, in fact, be more generative of relations of becoming-with. These alternatives, I contend, can be cultivated through playful tinkering.

Risk, detachment and alternative forms of togetherness

Detachment is regularly used to denote a sense of being 'one-step-removed' capable of facilitating an objective viewpoint (Ginn, 2014) or is presented as the negative opposite to attachment that manifests in unwillingness or inability to connect. However, such renderings fail to communicate the complexity of these relations. Candea demonstrates in his writing on human-animal (meerkat) relations, that attachment (or, in his words, engagement) and detachment should be seen '...not as a dichotomy but, rather, as a symbiosis: the vital, necessary, ever-changing, and often microscopic co-implication of two profoundly different forms' (Candea, 2010, p. 255). The nuance and co-production of such relations, he suggests, lays the foundations for new ways of conceiving of ethics that pushes at the boundaries of what it means to be with others in the world where tensions and harmony may be in constant flux in response to ongoing relational negotiations. While Candea specifically links this to an ethics of anthropology and focuses on the co-production of attachment and detachment, Ginn, in his discussion of London gardeners and their awkward relationship with slugs, sets his sights on detachment itself as an ethical basis.

For Ginn, 'the inevitability of detachment, not the fact of our being related, is a grounds for ethics' (Ginn, 2014, p. 538). He contends that 'dreams of detachment' (Ginn, 2014, p. 541) and accompanying desires to avoid sticky, mutually composing (or decomposing) encounters present as the basis for ethical relations through their capacity to expose mutual vulnerabilities among entities unknowable to each other yet who must live together:

> ...fleeting awareness of the irretrievability of the lives of others intensifies poignancy, such that despite a gulf separating the gardener from other creatures, some connection, however fleeting, is made to something—however strange. In this sense it is the space beyond relation that gives rise to inter-species ethics.
>
> (Ginn, 2014, p. 541)

In the moment of the encounter, there is recognition of alterity-in-relation (Ginn, Beisel & Barua, 2014), or alterity beyond relation which, when these events unfold with 'unloveable' creatures, induces a longed-for detachment which may manifest in a desire to cement our difference through distance and death. For Ginn, such encounters should not be glossed over or ignored in favour of conventional relations of care as the tensions in these spaces may provide openings for ethical reconfigurations. In the gardens he focuses on, Ginn finds that such encounters often invoke human benevolence that spares the slug. Such ethical reconfigurations are promising for supporting alternative conceptions of the Anthropos but this focus on detachment may also maintain attention on bounded subjects, where the human actor remains dominant.

Forms of togetherness which only draw on attachments and detachments and focus on life as unique to bounded entities may not offer a sufficient ethico-political impetus for more generative human-nonhuman relations. An opening up to the relational flows that enable bounded forms to take shape and survive—thus a recognition of the lively potential of togetherness-in-relation—may be critical for laying the foundations for how we can live well together. Attention to these relations could, as Brice writes, enact a 'thoroughgoing spatio-ontological reassembling' in which life becomes 'a vector of relation and recombination' (Brice, 2014, p. 180). Such reassemblages, where life exists in 'flows of energy and materials' (Brice, 2014, p. 186) among bounded entities, emphasises the outcome of relations rather than the mode themselves. Thus, the focus becomes less on notions of attachment or detachment and instead on how forms of togetherness are generated and their affective force.

Abrahamsson & Bertoni find something similar in their explorations of decomposing with worms, noting 'Vermicomposting is about doing togetherness in a way that is neither detached nor engaged' (Abrahamsson & Bertoni, 2014, p. 126). For these authors, worm farming manifests as a form of compost politics, a play on Stengers' notion of 'cosmopolitics', in which the players and their relations are regularly unknowable due to their complex, variable and contingent processes (2014, p. 133). As they observe:

> . . .you may not know, but rather become attuned to your worms. Compost politics is neither assimilation through identity nor the dream of harmony but rather a mutual domestication of multiple and different activities.
>
> (Abrahamsson & Bertoni, 2014, p. 134)

These are ways of being and doing not able to be simply conceptualised as attachments or detachments but instead best understood as interactions or extra-relational forms of togetherness. These embodied, active, responsive encounters speak to the notion of playful tinkering I develop here. One does not need to 'know' the other to play and play does not need to lead to some new, improved form of understanding or relational harmony. Play itself is primarily about the relations that unfold through recognition of mutual vulnerabilities and their generative capacity rather than the individual entities involved. Recognition of togetherness-in-relation is core to play.

Massumi talks in a similar way when he suggests that play enables 'mutual inclusion' where a holding apart, or difference, is maintained through modes of being and doing that induce an 'instantaneous back-and-forth. . .between the present and futurity' (2014, p. 23). This is not the hyper-separation of humans from nature that dominates anthropocentric visions but a separation predicated on recognition of difference and the capacity for learning to be affected by things we cannot possibly know and the productive potential of the resulting relations. These unknowable, propositional qualities mark out play as a process predicated on 'real' consequences of risky togetherness. These are risky because entities (including ourselves) with which we might engage or respond to can no longer be considered to be knowable—'smooth' and singular. Instead they are 'tangled objects,' bound up in multiples (Latour, 2004, p. 22) and configured through shifting relational interactions. To enter into such encounters requires: 'A daring not to be in complete control, or even to really know the other but to play anyway, . . .[and] a sense of humour' (Rose et al., 2011, p. 336). This daring may be key to the development of attuned multispecies relations as Rose et al. ask,

> With all of this in mind, perhaps the relevant question for multispecies relations is: Can people play? Do people have the skill, daring, humour and willingness to give themselves over to indeterminacy and the joy of encounter?
>
> (Rose et al., 2011, p. 336)

Throughout this text, we will meet people who do indeed exhibit the 'skill, daring and humour' to enter into playful encounters marked by forms of togetherness-in-relation. I hesitate to add 'willingness' to this list because these encounters are not always desired nor directly sought out. Indeed, engaging with the riskiness of play invites the potential of entering into encounters that are cognitively and kinaesthetically unsettling, unknown and unknowable and that may well subvert notions of free will. Such playful encounters are marked by risk and can be fraught with tension often experienced through the joy of attachment and sometimes a profound desire for 'detachment', but the 'doings' effects and affects of whatever mode of relations are engaged in constitute forms of togetherness which can be generative of alternative modes of being and doing. Thus, playful tinkering, as I argue throughout the text, presents as a means of developing the creative skills and capacities, or 'training', we will need to survive in our uncertain futures.

Training for uncertainty with playful tinkering

Risk and response-ability in the face of difference, of not knowing and being unable to know, are key to the generative potential of play. Haraway sees this as intimately linked to the 'roots of ethical possibilities... in play' (2014, p. 260) explored by biologist Mark Bekoff. Bekoff and his colleagues contend that play is sought out by the mammals that they study as a means of 'training for the unexpected' and this regularly includes risky or self-hindering play (see Spinka, Newberry & Bekoff, 2001). Such training leads to what we might call resilience, though the appropriation of such a word in policy no doubt makes many of us wary. At the very least, we could say that Spinka et al.'s research indicates that for the mammals they refer to, play has the potential to enhance the skills and capacities cognitively, kinaesthetically and emotionally of players who engage in the associated risks. Similar claims are made about the role of play in children's development.

It is the very riskiness and uncertainty of play accompanied by its position as an active mode of doing capable of leading to ontological reimaginings of the subject and the world within which the subjects come to be that differentiates this concept from Bennett's notion of naive realism as a means of being open to forms of thing-power and the relations through which this becomes configured.

While naive realism invites many risks induced by exposing the vulnerabilities of humans in relation to the limits of control over matter, it is not necessarily an engagement entered into where risk is intentionally induced through new modes of doing. Engaging with the riskiness of play, on the other hand, invites the potential of entering into encounters that are cognitively and kinaesthetically unsettling, unknown and unknowable as a means of 'training' for contingent futures predicated on notions of becoming-with. As Haraway writes,

> Play always involves the invitation that asks 'are we a "we"'? A 'we' that doesn't pre-exist the propositional risk and testing. I think all the important problems involve this propositional, questioning, interrogative 'we'.
>
> (2014, p. 261)

In the modes of play explored by Bekoff, (2007) (and Haraway's restorying of these), participants tend to be humans or our not too distant relatives rather than inert objects or living entities whose presences are, perhaps, less apparently lively (see Puig, 2010 for a beautiful discussion of shifts in understandings of soil to position it as 'lively'). However, Shields argues that there are elements of play 'existing at least partially outside the human experience of it' (2015, p. 299). I would suggest this could be extended to include also being outside of the mammalian experiences explored in Bekoff's and Splinka et al.'s work and also that of Bateson (1972) and Massumi (2014). Indeed, Massumi suggests an expansion of the experience of play in his discussion of instinct and intuition in relation to Darwin's observations of earthworms and their capacity to invent and vary actions in response to, or in concert with, the contingencies of their unique contexts. He also hints at the potential to extend this further when he writes '[t]he plant participates in animality and vice versa' (Massumi, 2014, p. 52), but he appears to reign this in when later referring to animal 'playfulness' seemingly in juxtaposition to plant 'modesty' (2014, p. 54). However, in the reconfiguration of notions of subjectivity through 'plantiness' in the work of Head et al. (2015), and through the shifts and adaptations that occur in the plant world in response to environmental changes, we may also glimpse the basis for conceptions of planty forms of play.

Uncertainty is a hallmark of play, and ongoing engagement in play in a tinkering mode is a means of training our skills and capacities to attune and respond to the unexpected and, sometimes, this is generative of new modes of being and interaction. As Henricks writes: 'playful behavior is . . .a protest against orders and orderliness' (2006, p. 209). Haraway echoes this when she observes that the joy of play 'breaks rules to make something else happen' (2008a, p. 459) and Shields notes play can induce:

> . . .the momentary, unsettling experience of feeling the world in a wholly unknown, indescribable way is the avenue through which dominant logics are interrupted and the boundaries of language—not as immovable as perhaps has been thought—may actually be transcended.
>
> (2015, p. 316)

The embodied, affective realm of play is propositional. 'Play makes an opening. Play proposes' (2008b, p. 240). In this 'contact zone' (Haraway, 2008b), play is generative. It is the doing, the becoming-with in play, that enables this. While Shields focuses on this as enabling the transcendence of language to emphasise the potential of play to move beyond accepted logic, I suggest that these skills and capacities have the material-semiotic force capable of conceiving and enacting alternative worlds capable of supporting more sustainable modes of being and doing.

Rose et al., (2011) suggests that the potentiality of play can be conceived of as a 'charge'. Charge in the sense drawn on here focuses on the 'the dynamic relation between a myriad of charged particles' (Rose et al., 2011, p. 334). Each entity entering into the material-semiotic field may have its own charge, but it is in their interplay—their responsive interactions—that play both enacts and is enacted. As such, as Rose et al., write 'play does not only happen within a field of interactions, in an important sense it is also a charge which generates that field and the relationships that comprise it' (2011, p. 337). Play exceeds the sum of its parts and is itself 'a surplus: an excess of energy or spirit' (Massumi, 2014, p. 9) capable of generating creative reimaginings. It may be that we need to train our skills and capacities to enter into, and to enact play in a tinkering mode, to ignite the charge capable of challenging anthropocentric modes of world-making centred on resource-draining 'exceptional' human subjects. If we accept that play is not just the form taken by particular sets of practices but a 'charge', it could also be a means of reshaping fields of interactions and the encounters that form it (Rose et al., 2011). Here, play is positioned not simply as a means of developing new relational modes of acting in the world, but capable of igniting the generative possibilities of developing new worlds predicated on responsive, entangled relationships operationalised in ongoing play.

Play invites engagement in, and training for, ontological choreography (Haraway, 2008b) that has the capacity to shift ingrained patterns of behaviours and thought and incite 'creative destabilizing action' (Schechner, 1988, p. 17). It provides opportunities to exceed the bodies overtly engaged in the choreography to become more than the sum of their parts while being induced by these very visceral encounters. The concept of convivial dignity is developed in the final section of this chapter as a way of conceptualising these 'more-than' encounters that rely on multifarious forms of togetherness-in-relation. I contend that playful tinkering is best able to support attunement to uncertainty and invocation of the adjustments needed to support sustainable futures when enacted with an ethico-political notion of convivial dignity that seeks to privilege the relational aspects of togetherness.

Understanding convivial dignity

Convivial dignity provides a way of talking about relations that are not predicated on assumptions of attachment, detachment nor mobilised by notions of care (which I argue throughout this text tend to reproduce anthropocentric modes of thinking). Instead, they are grounded in necessary togetherness and recognition that mutual survival is induced by and depends on these relational interactions. The use of convivial takes its cue from Hinchliffe and Whatmore's notion of a 'politics of conviviality' which they state 'is serious about the heterogeneous company and messy business of living together' (2006, p. 134). It is employed here to refer to recognition of the inescapability of our togetherness in alterity and is a term commonly encountered in work attending to the more-than-human. On the other hand, my use of dignity is an attempt to draw on the term's affective force while simultaneously speaking back to the dominant ways in which it tends to be conceptualised. I do this by extending its purview beyond both human and more-than-human bounded entities to the modes of encounter—or the very relations of togetherness—that unfold. Life then, in these configurations, is conceived of as flows and interdependencies rather than simply relations among bounded entities.

Convivial dignity is an attempt to provide a grammar capable of conceptualising ways in which our entangled existence exceeds the sums of its parts. In so doing, it aims to highlight the generative potential of these relations as well as to support and amplify our material-semiotic engagement in these. Overarchingly, the grammar of convivial dignity attempts to contribute to the semiotic toolbox for reimagining the human exceptionalism commonly identified as defining the Anthropos at the heart of the Anthropocene. In so doing, convivial dignity represents an attempt to support less resource intensive and exploitative modes of being and doing with the world through the suggestion of alternative ontologies. The enactment of convivial dignity is suggested to be a hallmark of forms of playful tinkering that are generative of productive forms of togetherness-in-relation.

To flesh out the potential of this new grammar, it is important that I first dwell with the notion of dignity. While I use convivial in a rather conventional manner, my recasting of dignity is a much riskier move. So, I ask you to be open to the surprise such risky endeavours can elicit as we undertake efforts to imagine better futures. Part of this risk is induced by the flexibility of the term dignity itself, as attested to by ongoing debates about its meaning and utility within political theory. There within, positions range across the spectrum from those asserting that the term is an unhelpful distraction through to those who assert that it is fundamental to our conception of humanity, and consequently, human rights. Dignity has perhaps most commonly been employed to separate humans from animals with little to no regard for other nonhumans. As already identified, this text seeks to appropriate the term to intervene in the grammar and rhetoric underpinning and fuelling anthropocentric practices and beliefs. To flesh out the rationale for this manoeuvre, I first provide a brief sketch of key philosophical conceptions and political uses of dignity to explore the ways in which its humanist framework has been and can be extended to the more-than-human and the relations through which we can enact and narrativise the necessity of living together.

Human dignity, human exceptionalism and crisis

Dignity has a history of being called upon in times of crisis. Talk of human dignity gained momentum in post WW2 international efforts to grapple with ways of conceptualising the atrocities inflicted on some humans and to craft narratives (as well as agreements, accords and laws) to avoid repeating them. The use of dignity in this text also emerges in response to crisis and atrocity at the multispecies and more-than-human scale and is used to explore ways of attuning practices and subjects to the hope and grief (Head, 2016) of the Anthropocene. But shifting the discursive purview of dignity beyond humans is not an easy task. The modern conception of dignity is typically embedded with assumptions about individual autonomy and free will. These are assumptions often linked to the work of Kant, though these are not associations he himself made.

For Kant, human autonomy conceived of as the capacity to exercise morality constitutes 'the ground of the dignity of human nature' (Kant, 1998, 4:436). This does not, however, equate to a notion of human capacity to act without regard for all else as if we are laws unto ourselves. Rather, as Rosen writes, '[w]hat Kant has in mind as autonomy is the idea that the moral law which we must acknowledge as binding upon us is "self-given"' (Rosen, 2012, p. 25). Dignity, thus, resides in the law, but this is then brought into practice through human action. Nonetheless, human exceptionalism pervades contemporary conceptions of dignity commonly linked back to Kant. However, as Anderson has demonstrated through her discussion of 'anatomical humanism', the perpetuation of such ideas is not simply a result of metaphysical logic but is also deeply embedded in renderings of bodily matter (Anderson, 2014).

The International Human Rights conception of the inalienable 'dignity and worth of the human person' (preamble) and the assertion 'All human beings are born free and equal in dignity and human rights' (UN, 1948, Art 1) are perhaps the most well-known manifestations of this materially supported conception of human exceptionalism. Of course, despite these proclamations, dignity is not always thought about in relation to equality among this one species. Variation of moral duties and obligations and thus rights can be seen in numerous instances including attempts to develop intra-species hierarchies through craniology and its links to race and ethnicity and also in conceptions of those with disabilities, particularly in relation to perceived mental capacity. The materially informed perpetuation of these ideas is evident also in the most commonly identified violations of human dignity in human rights that centre on bounded, individual bodies. Slavery, starvation, torture, rape all emphasise the vulnerability of bodies but levels of vulnerability are not shared equally within and among species. Through the violation of bodies, we see that the narratives and practices of anthropocentric forms of dignity are not simply maintained by Cartesian metaphysics but through particular forms of attention to matter.

While I am glossing over much of the debate in this area, my aim is simply to unsettle the humanist conception of dignity by drawing attention to its metaphysical and material manifestations. In so doing, I wish to reinforce the argument that radical changes to anthropocentric thinking require more than a material turn. The material is already imbricated in practices of human exceptionalism, as evident in the uses and attributions of dignity. These modern renderings of dignity makes it a risky term to draw on in work attempting to subvert human exceptionalism. Yet, it is the affective force of this concept that makes it ripe for reworking to explore how it could be mobilised to support and build more responsive, attuned relations in contemporary and future experiences of uncertainty. If, as Habermas would have it, human dignity 'feeds off the outrage of the humiliated at the violation of their human dignity' (2012, p. 75) then the affective force of its material, bodily manifestations beyond bounded human entities, are worth investigating. The very slipperiness of the term leaves it open for play. Martha Nussbaum, in contrast to the majority of liberal political theorists, has seized upon this slipperiness in the development of her capabilities view where she attempts to broaden the social justice scope of dignity to include animals. The uniqueness of this approach deserves special attention here.

Dignity, capabilities and mutual vulnerabilities

Nussbaum outlines ten capabilities that enable a dignified life to be experienced. These are: 'Life... Bodily Health... Bodily Integrity... Senses... Imagination... Thought... Emotions... Practical Reason... Affiliation... Other Species... Play... [and] Control Over One's Environment' (Nussbaum, 2006, pp. 76–77). Drawing on an Aristotelian notion of flourishing, Nussbaum argues for a plural approach to dignity noting 'it must never be forgotten that this [human dignity] is just one type of dignity that laws and institutions should recognize' (Nussbaum, 2011, p. 28). Indeed, she goes on to assert:

> We need an expanded notion of dignity since we now need to talk not only about lives in accordance with human dignity but also about lives that are worthy of the dignity of a wide range of sentient creatures ...The Capability Approach regards each type of animal as having a dignity all its own...The species plays a role in giving us a sense of the characteristic form of life that ought to be promoted.
>
> (Nussbaum, 2011, p. 161)

The Capability Approach challenges traditional interpretations of social contract theory by aiming to bring people with disabilities, people across nation-states and nonhuman animals into the fold of social justice. While this more open approach may allow for multiple dignities to exist, it offers a limited conception of the social. This is most evident in Nussbaum's eighth capability which focuses on 'other species: Being able to live with concern for and in relation to animals, plants, and the world of nature' (2006, p. 77). While a 'holding apart' of humans and nonhumans in regards to acknowledgement of alterity, as can be read into this capability, is not necessarily problematic, her conception tends to reinforce forms of hyper separation where interspecies relations are assumed to be quite static.

> The privilege of human exceptionalism also appears to persist in Nussbaum's concern with the discrete, bounded entities that make-up the social rather than with the productive capacity of the relational encounters and flows among these. As Holland and Linch observe, while the capabilities approach provides 'a basis for securing the environmental preconditions of human flourishing as a fundamental entitlement, and for extending this support to include the specific excellence or dignity of non-human life' (2016, p. 413), it does not 'specify the theory's implications for understanding the role of environmental relationships in human and non-human flourishing and for establishing how to balance, prioritize, and make trade-offs between the capabilities of different species' (Holland & Linch, 2016, p. 416). Efforts to analyse the implications of the capabilities approach have prompted arguments that support anthropocentric protections of the environment as well as attempts to justify radically new conceptions of justice for nonhumans.

The notion of convivial dignity developed here attempts to push at the very make-up and boundaries of how we conceive of the social, specifically challenging its bracketing off from more-than-humans and the persistence of concern with bounded entities. Nussbaum's attention to nonhumans manifests in the extension of dignity to animals but the messy relations of convivial dignity (brought about by its attention to the unknowable outcomes of human/nonhuman relational entanglements) pushes this further to encompass other nonhuman entities, relations and their effects.

In so doing, convivial dignity emphasises the mutual vulnerabilities that are brought to the fore when we recognise the necessity of our togetherness-in-relation. For Hird, (2013), such exposure lays the groundwork for an ethics of vulnerability that is not predicated on intimacy, shared recognition or empathy but on the necessity of our being together in sometimes unsettling mutual entanglements with entities often unknown and unknowable to each other. Such a position is predicated on an ethics of responsibility, one embedded in noticing and attuning, one that, at its core, is about humans encountering relations that are not only beyond their control but micro-ontologies where humans do not figure at all. Vulnerability is not simply about being exposed or susceptible to risk and thus in need of special forms of care and consideration; It is also a 'condition of receptivity' (Green & Ginn, 2014, p. 152). Indeed, Green and Ginn argue that recognition of vulnerability can induce action by encouraging the cultivation of practices that aim to reshape the present with an eye to the future.

Recognition of mutual vulnerabilities, such as in the alternative beekeeping Green and Ginn study, provides 'the transformative, ethical heart' (Green & Ginn, 2014, p. 164) of the practice. While these vulnerabilities (such as, in this example, the bees being vulnerable to the caring practices of the keepers and the keepers to the sting of the bees) may not be shared equally among human and more-than-humans, ongoing exposure to these potentially awkward encounters is shown to facilitate generative ethical meetings capable of reshaping human/nonhuman relations. Mutual relations of vulnerability can induce experiences that support recognition and enactment of dignity in a convivial manner that extends its purview beyond bounded human and more-than-human entities to the very relational flows that enliven these entities. This conception of convivial dignity aims to offer a new grammar capable of initiating a 'linguistic jolt' (Buck, 2013) that unsettles the humanist subject and expands the purview of our concern to relations-in-togetherness. In so doing, it attempts to capitalise on the receptive characteristics induced by recognition of mutual vulnerabilities and the affective force of notions of dignity to encourage forms of relational entanglements best able to promote generative and creative practices that challenge anthropocentrism.

Conclusion

This chapter has taken the form of a degustation menu. Whilst ranging across numerous fields and concepts, I have attempted to offer just enough of each course to ensure we are sated but not overstuffed before we get to the main event. In so doing, there is much I have glossed over and so much more to say. But, in the very least, I hope this appetiser has whet your appetite. We began this chapter by attending to the metabolic body. In so doing, I drew attention to the ways the eating body highlights the very fragility of the hyper-separation of the humanist Anthropos that sits at the heart of dominant conceptions of the Anthropocene. Eating, as we have seen, cannot be done by the human subject alone. As Tsing writes, 'we eat for ourselves and others' (2014, p. 28). At one level, this refers to the millions of companions that inhabit—or perhaps more precisely enact—our (their?) bodies.

When humans eat, the flora and fauna of nonhuman bodies are also nourished. Human bodies, then, are sites of multiple forms of feedings and flourishing, some of which can threaten the life of the body itself. This does not sit well with the master narrative of humanism. The necessity of counteracting these dominant narratives, however, is shown to require more than a material-turn. As we have seen, the material has also been enrolled in justifications of human exceptionalism. The challenge then, requires material-semiotic responses. We not only need new ways of being and doing with the world but alternative grammars that can conceptualise and amplify these. To attend to these demands throughout this book I experiment with the generative potential of concepts of play and convivial dignity to demonstrate how alternative ontologies and world-making practices could be mobilised.

Playful tinkering, as a variant of experimentation, has been shown to be capable of inducing recognition of the inescapability of togetherness-in-relation. Play requires an openness to being moved by and moving with unknown and unknowable others. Through the complexity of these movements, play can be conceived of as enabling 'ontological choreography' (Haraway, 2008b) among humans, more-than-humans and beyond these bounded entities to the very relational flows that enable these configurations to take shape. Its very openness makes play propositional in nature. We don't know what the outcome of these complex, shifting and necessarily responsive relations will be. This non-teleological characteristic makes play a risky venture, one that can be conceptualised as a form of composting; as a process of decomposition rather than composition. Here, decomposing is understood not simply as a site of destruction but of transformation—of relations where detachment is often central but togetherness is required. While composition infers a positive creation and sustaining of particular forms of life, the messy, stinky engagement with death and decay of decomposition is suggestive of much less 'cosy' relations often involving 'unloved' (Rose, Cooke & van Dooren, 2011) and 'Monstrous' (Ginn, 2014) others.

Indeed, play is a form of generative interaction and a charge capable of contributing to the making of new worlds that can be enacted without an affect of love or ethic of care. It doesn't require familiarity or forms of attachment. In fact, detachment can be useful here insofar as it draws attention to alterity-in-relation (Ginn et al., 2014). The very material flows of decomposition are sites where the unknown and unknowable relations and flows of life are viscerally apparent. In this way, playful tinkering provides opportunities for training to become attuned and responsive to shifting relations with strange strangers (Morton, 2010). In so doing, engagement in play provides opportunities for players to develop their skills and capacities to adopt openness and adapt to uncertainty.

These skills and capacities are most effectively geared towards challenging anthropocentric modes of being and doing when they are guided by convivial dignity. Convivial dignity is offered as an alternative grammar that intervenes in normative narratives of human exceptionalism by functioning as an ethico-political guide for forms of play that attempts to respond to Gibson-Graham's question, 'What do humans, other species and ecosystems need in order to survive with some kind of dignity?' (Gibson-Graham & Roelvink, 2010, pp. 333–334). While Gibson-Graham suggest this is an 'anthropomorphic question', I contend that dignity can be stretched outside of these humanist frameworks to more-than-humans and the very relational flows that enliven entities and ignite play.

Convivial dignity offers a guide for responsive, adjusting, playful tinkerings attuned to the relational flows and interdependencies through which life is enacted. This is not predicated on the pursuit of justice, but on attunement to mutual vulnerabilities. Convivial dignity is responsive to the irreducible, recalcitrant characteristics of more-than-humans and the relations through which these are formed. It is the generative, enacting nature of these relations—the forms of togetherness and becoming-with—that are of key concern in convivial dignity. This recognition of vulnerability through the togetherness-in-relation of playful tinkering guided by convivial dignity can encourage alternative ways of being and doing and perhaps the building of more ecologically attuned worlds.

To explore just how these material-semiotic concepts might function in practice, the rest of this book engages with embodied, visceral encounters with food. As we will see, these encounters will not always be harmonious or positive experiences. There will be death, tension and fear that seem at odds with the potential for these ideas and their associated actions and narrative force to support recognition of entangled lives attuned and responsive to uncertainty. But relational lives are messy. The gardeners, exhibitors, shoppers and composters we will meet offer the chance to explore this messiness, contemplate what else might be possible and consider ways to cultivate and stretch these potentialities to encourage local action in the face of uncertain futures. Food, from taste to waste, provides us with ample opportunities for engagement with the propositional potential of play and, when guided by convivial dignity, can support alternative ways of being and doing in the world. It is just this willingness to encounter risk and be open to surprise that could amplify a 'feeling life' (Thrift, 2000) and cultivate more sustainable ways of eating for and with our future.

References

Abrahamsson, S., & Bertoni, F. (2014). Compost politics: Experimenting with togetherness in vermicomposting. *Environmental Humanities*, *4*, 125–148.

Abrahamsson, S., Bertoni, F., Mol, A., & Ibáñez Martín, R. (2015). Living with omega-3: New materialism and enduring concerns. *Environment and Planning D: Society and Space*, *33*(1), 4–19.

Allon, F., & Sofoulis, Z. (2006). Everyday water: Cultures in transition. *Australian Geographer*, *37*(1), 45–55.

Anderson, K. (2007). *Race and the crisis of humanism*. London & New York: Routledge.

Anderson, K. (2014). Mind over matter? On decentring the human in human geography. *Cultural Geographies*, *21*(1), 3–18.

Atchison, J., & Head, L. (2013). Eradicating bodies in invasive plant management. *Environment and Planning D: Society and Space*, *31*(6), 951–968.

Bateson, G. (1972). *Steps to an ecology of mind*. Chicago, IL: The University of Chicago Press.

Bawaka, C., Wright, S., Suchet-Pearson, S., Lloyd, K., Burarrwanga, L., Ganambarr, R., . . . Sweeney, J. (2015). Co-becoming Bawaka: Towards a relational understanding of place/space. *Progress in Human Geography*, *40*(4), 455–475.

Bekoff, M. (2007). *The emotional lives of animals*. Novato, CA: New World Library.

Bennett, J. (2010). *Vibrant matter: A political ecology of things*. Durham, NC: Duke University Press.

Brice, J. (2014). Killing in more-than-human Spaces: Pasteurisation, Fungi, and the Metabolic Lives of Wine. *Environmental Humanities*, *4*(1), 171–194.

Buck, H. J. (2013). Climate engineering: Spectacle, tragedy or solutions? A content analysis of news media framing. In C. Methmann, D. Rothe, & B. Stephan (Eds.), *Interpretive approaches to global climate governance: Deconstructing the greenhouse* (pp. 166–181). New York: Routledge.

Cameron, J. (2015). On experimentation. In K. Gibson, D. B. Rose, & R. Fincher (Eds.), *Manifesto for living in the Anthropocene*. New York: Punctum Books.

Candea, M. (2010). "I fell in love with Carlos the meerkat": Engagement and detachment in human–Animal relations. *American Ethnologist*, *37*(2), 241–258.

Derrida, J. (1991). Eating well, or, the calculation of the subject: An interview with Jacques Derrida. In E. Cadava, P. Connor, & J. L. Nancy (Eds.), *Who comes after the subject?* (pp. 98–119). New York: Routledge.

Derrida, J., & Caputo, J. D. (1997). *Deconstruction in a nutshell: A conversation with Jacques Derrida*. New York: Fordham University Press.

Dibley, B. (2012). "The shape of things to come": Seven theses on the Anthropocene and attachment. *Australian Humanities Review*, *52*, 139–158.

Gibson-Graham, J. K. (2008). Diverse economies: Performative practices for other worlds. *Progress in Human Geography*, *32*(5), 613–632.

Gibson-Graham, J. K. (2011). A feminist project of belonging for the Anthropocene. *Gender, Place and Culture*, *18*(1), 1–21.

Gibson-Graham, J. K., & Roelvink, G. (2010). A economics ethics for the Anthropocene. *Antipode: A Radical Journal of Geography*, *41*, 320–346.

Ginn, F. (2014). Sticky lives: Slugs, detachment and more-than-human ethics in the garden. *Transactions of the Institute of British Geographers*, *39*(4), 532–544.

Ginn, F. (2015). When horses won't eat: Apocalypse and the Anthropocene. *Annals of the Association of American Geographers*, *105*(2), 351–359.

Ginn, F., Beisel, U., & Barua, M. (2014). Flourishing with awkward creatures: Togetherness, vulnerability, killing. *Environmental Humanities*, *4*(1), 113–123.

Green, K., & Ginn, F. (2014). The smell of selfless love: Sharing vulnerability with bees in alternative apiculture. *Environmental Humanities*, *4*(1), 149–170.

Habermas, J. (2012). *The crisis of the European Union: A response*. Cambridge: Polity Press.

Haraway, D. (1997). *Second_Millennium.FemaleMan_Meets_OncoMouse Feminism and Technoscience*. New York and London: Routledge.

Haraway, D. (2008a). Training in the contact zone: Power, play and invention in the sport of agility. In B. Da Costa & K. Philip (Eds.), *Tactical biopolitics: Art, activism and technoscience* (pp. 445–464). Cambridge, MA: The MIT Press.

Haraway, D. (2008b). *When species meet*. Minneapolis & London: University of Minnesota Press.

Haraway, D. (2014). Anthropocene, capitalocene, chthulhocene. Donna Haraway in conversation with Martha Kenney. In H. Davis & E. Turpin (Eds.), *Art in the Anthropocene: Encounters among aesthetics, politics, environments and epistemologies* (pp. 255–270). London: Open Humanities Press.

Haraway, D. (2015). Anthropocene, capitalocene, plantationocene, chthulucene: Making kin. *Environmental Humanities*, *6*(1), 159–165.

Harbers, H., Mol, A., & Stollmeyer, A. (2002). Food matters: Arguments for an ethnography of daily care. *Theory, Culture & Society*, *19*(5–6), 207–226.

Hayes-Conroy, J., & Hayes-Conroy, A. (2013). Veggies and visceralities: A political ecology of food and feeling. *Emotion, Space and Society*, *6*, 81–90.

Head, L. (2008). Nature, networks and desire: Changing cultures of water in Australia. In P. Troy (Ed.), *Troubled waters: Confronting the water crisis in Australia's cities* (pp. 67–80). Canberra: ANU E Press.

Head, L. (2014). Contingencies of the Anthropocene: Lessons from the 'Neolithic'. *The Anthropocene Review*, *1*(2), 113–125.

Head, L. (2016). *Hope and grief in the Anthropocene: Re-conceptualising human–nature relations*. London & New York: Routledge.

Head, L., Atchison, J., & Phillips, C. (2015). The distinctive capacities of plants: Rethinking difference via invasive species. *Transactions of the Institute of British Geographers*, *40*(3), 399–413.
Henricks, T. S. (2006). *Play reconsidered: Sociological perspectives on human expression*. IL: University of Illinois Press.
Heuts, F., & Mol, A. (2013). What is a good tomato? A case of valuing in practice. *Valuation Studies*, *1*(2), 125–146.
Hinchliffe, S., & Whatmore, S. (2006). Living cities: Towards a politics of conviviality. *Science as Culture*, *15*(2), 123–138.
Hird, M. J. (2013). Waste, landfills, and an environmental ethics of vulnerability. *Ethics and the Environment*, *18*(1), 105–124.
Holland, B., & Linch, A. (2016). Cultivating human and non-human capabilities for mutual flourishing. In T. Gabrielson, C. Hall, J. M. Meyer, & D. Schlosberg (Eds.), *The Oxford handbook of environmental political theory* (pp. 413–428). Oxford: Oxford University Press.
Kant, I. (1998). *Groundwork of the metaphysics of morals* (M. J. Gregor, Trans.). Oxford and New York: Oxford University Press.
Kristeva, J. (1982). *Powers of horror: An essay on abjection* (L. S. Roudiez, Trans.). New York: Columbia University Press.
Latour, B. (2004). How to talk about the body? The normative dimension of science studies. *Body & Society*, *10*(2–3), 205–229.
Law, J., & Mol, A. (2002). Complexities: An introduction. In J. Law & A. Mol (Eds.), *Complexities: Social studies of knowledge practices* (pp. 1–22). Durham, NC: Duke University Press.
Law, J., & Mol, A. (2008). The actor-enacted: Cumbrian sheep in 2001. In C. Knappett & L. Malafouris (Eds.), *Material agency: Towards a non-anthropocentric approach* (pp. 57–78). Dusseldorf: Springer.
Lorimer, J., & Driessen, C. (2014). Wild experiments at the Oostvaardersplassen: Rethinking environmentalism in the Anthropocene. *Transactions of the Institute of British Geographers*, *39*(2), 169–181.
Malm, A., & Hornborg, A. (2014). The geology of mankind? A critique of the Anthropocene narrative. *The Anthropocene Review*, *1*(1), 62–69.
Massumi, B. (2014). *What animals teach us about politics*. Durham and London: Duke University Press.
Mol, A. (2008). I eat an apple. On theorizing subjectivities. *Subjectivity*, *22*(1), 28–37.
Mol, A. (2009). Good taste: The embodied normativity of the consumer-citizen. *Journal of Cultural Economy*, *2*(3), 269–283.
Mol, A. (2010a). Actor-network theory: Sensitive terms and enduring tensions. *Kölner Zeitschrift für Soziologie und Sozialpsychologie Sonderhefte*, *50*(1), 253–269.
Mol, A. (2010b). Care and its values: Good food in the nursing home. In A. Mol, I. Moser, & J. Pols (Eds.), *Care in practice: On tinkering in clinics, homes and farms* (pp. 215–234). Bielefeld, Germany: Transcript Verlag.
Mol, A., Moser, I., & Pols, J. (2010). Care: Putting practice into theory. In A. Mol, I. Moser, & J. Pols (Eds.), *Care in practice: On tinkering in clinics, homes and farms* (pp. 7–26). Bielefeld, Germany: Transcript Verlag.
Moore, J. W. (2017). *The Capitalocene, Part I: On the nature and origins of our ecological crisis*. The Journal of Peasant Studies, *44*(3), 594–630.
Morton, T. (2010). *The ecological thought*. Cambridge, MA: Harvard University Press.

Nussbaum, M. (2006). *Frontiers of justice: Disability, nationality, species membership.* Cambridge, MA: Harvard University Press.

Nussbaum, M. (2011). *Creating capabilities: The human development approach.* Cambridge, MA: Harvard University Press.

Plumwood, V. (1996). Being prey. *Terra Nova, 1*(3), 33–44.

Plumwood, V. (2007). A review of Deborah Bird Roses's 'reports from a wild country: Ethics for decolonisation'. *Australian Humanities Review, 42.*

Probyn, E. (2012). In the interests of taste and place: Economies of attachment. In V. Rosner & G. Pratt (Eds.), *The Global and the intimate: Feminism in our time* (pp. 57–84). New York: Columbia University Press.

Puig De La Bellacasa, M. (2010). Ethical doings in naturecultures. *Ethics, Policy & Environment. A Journal of Philosophy and Geography, 13*(2), 151–169.

Robbins, P., & Moore, S. A. (2013). Ecological anxiety disorder: Diagnosing the politics of the Anthropocene. *Cultural Geographies, 20*(1), 3–19.

Rose, D. B. (1999). Indigenous ecologies and an ethic of connection. In N. Low (Ed.), *Global ethics and environment* (pp. 175–187). London & New York: Routledge.

Rose, D. B., Cooke, S., & Van Dooren, T. (2011). Ravens at play. *Cultural Studies Review, 17*(2), 326–343.

Rosen, M. (2012). *Dignity: Its history and meaning.* Canbridge, MA: Harvard University Press.

Schechner, R. (1988). Victor Turner's last adventure. In V. W. Turner (Ed.), *The Anthropology of performance* (pp. 7–20). New York: PAJ Publications.

Shields, R. (2015). Ludic ontology: Play's relationship to language, cultural forms, and transformative politics. *American Journal of Play, 7*(3), 298–321.

Spinka, M., Newberry, R. C., & Bekoff, M. (2001). Mammalian play: Training for the unexpected. *Quarterly Review of Biology, 76*(2), 141–168.

Stengers, I. (1997). *Power and invention: Situating science* (P. Bains, Trans.). Minneapolis: University of Minnesota Press.

Thrift, N. (2000). Still life in the nearly present time: The object of nature. *Body and Society, 6*(3–4), 34–57.

Tsing, A. (2014). More-than-Human sociality: A call for critical description. In K. Hastrap (Ed.), *Anthropology and nature* (pp. 27–42). New York: Routledge.

UN, (1948). *Universal Declaration of Human Rights,* http://www.un.org/en/universal-declaration-human-rights/

Weir, J. K. (2009). *Murray river country: An ecological dialogue with traditional owners.* Canberra: Aboriginal Studies Press.

Whatmore, S. J. (2006). Materialist returns: Practising cultural geography in and for a more-than-human world. *Cultural Geographies, 13*(4), 600–609.

Whatmore, S. J. (2013). Earthly powers and affective environments: An ontological politics of flood risk. *Theory, Culture & Society, 30*(7–8), 33–50.

Whatmore, S. J., & Landström, C. (2011). Flood apprentices: An exercise in making things public. *Economy and Society, 40*(4), 582–610.

3 Introducing Taste

Introduction

My childhood tasted like sweet, juicy bursts of sun-warmed strawberries. That and the crunch of just-pulled carrots rinsed free of the clinging black soil under the tap in my grandparents' garden. There were other tastes too, but these are the ones that stay with me. I wonder if they would taste the same now? My family's current garden has suffered in the blisteringly dry heat of the last weeks of summer, the absence of moisture benefitting only the garlic harvest. We hope we have enough to last the year. Today, I have returned from our local farmers' market with oodles of new season produce I found hard to resist. As summer gives way to a cooler and hopefully wetter autumn, apples, nashi pears and potatoes are again on offer. The supply of the once longed-for stone fruit is dwindling. There will only be a few more weeks of my favourite slightly tart, crisp plums and the blueberries that explode with indescribable flavour. But my tastes have turned now. I've really missed these autumnal foods.

This chapter takes the contradictory natures of taste as its focus to explore key debates about human and more-than-human relationships in the contemporary era of the Anthropocene as a precursor to exploration of the manifestations of taste encountered in the fieldwork in three different sites in the following three chapters. To lay these foundations, I first identify the anthropocentric modes of thinking invoked by linking taste to distinction (specifically in relation to food) and contrast these to the growing body of work that attempts to disrupt dominant forms of humanism marked by hyper-separation through a focus on the embodied, visceral, sensorial and, thus, necessarily relational experiences of taste. In so doing, we will see that taste, while sociological, also exceeds this categorisation with the very materiality of food and our visceral encounters with it enabling a simultaneous pushing at the very boundaries of what we have come to know as the social. Through taste, we encounter these excess, more-than moments via regular experiences with the recalcitrance of matter and the relations that enliven it/us.

Rather than being contained and knowable, taste will be shown to be produced through active, relational processes of becoming-with. In this text, the invocation of movement and being-in-process that explodes out of these becomings provides sustenance for the arguments that, if we are to enhance and develop the skills and

DOI: 10.4324/9780429424502-3

capacities needed to live in times of uncertainty, we must—and are capable of—enacting new ways of being and doing with the world that recast the contemporary humanist subject. This position does not dismiss taste as a form of social distinction but aims to contribute to an expansion of understandings by cultivating tastes for alternative flavours and textures. Embodied attunement to taste, as encountered through shifting relations among biological response, material attributes of foodstuffs and social, economic and environmental conditions, is shown to induce humans to act, offering opportunities to destabilise and re-materialise daily habits, behaviours and beliefs.

The relational becomings of taste and its potential affective force make it a particularly productive encounter through which to explore how we can 'be' and 'do' in ways that attempt to counteract the destructive modes of assumed human dominance and hyper-separation that characterise the Anthropocene. An openness to playing with tastes can prime us for engagement in responsive ontological choreographies capable of laying the foundations for alternative practices of world-making. Here in these Canberra households, we find this manifesting in approaches centred on agile forms of making do with the relational entanglements of gardening produce and AFN provisioned food. These are responses informed by recognition of the possibilities for the reconfiguration of relations among matter and of the very relations that bring particular materialities into view. This, I suggest, can underpin environmentally responsible ways of living even when people are not motivated by environmental sensibilities.

Throughout this chapter, I suggest that attuned modes of tasting can be conceived of as forms of playful tinkering that encourage development of the affordances required to recognise the necessity of togetherness-in-relation. This involves not only being moved by matter but being moved by the propositional and generative potential of the relations that enliven it. The capacity for reconceptualising attuned taste as a form of play capable of supporting creative reimaginings that move beyond dominant anthropocentric modes of being and living is enhanced when it is guided by convivial dignity. Convivial dignity provides a way of talking about taste as relational that is not predicated on assumptions of attachment or detachment, nor mobilised by notions of care. Instead, it is predicated on recognition of mutual vulnerabilities; what and how we taste is dependent on the multifarious relations that enable bodies and foods to congeal in particular, albeit fleeting, ways. The manner in which these tastes escape from normative conceptions of human mastery provides a fruitful basis for interrogation of alternative narratives capable of representing and amplifying these relational materialities and the ways of being and doing they can support and make possible.

Embodying taste

Taste is a hugely complex concept attended to in multiple ways in a wide variety of disciplines from sensory science to cultural theory. I do not attempt to canvas all of these here, instead offering a broad brushstrokes view of key concepts so I can focus my attention on understandings of taste emerging over recent years as part of the so-called 'affective' or 'visceral' turn (Goodman, 2016). Through these 'turns' a body of work has developed that critiques the narrow conception of taste accounted for by social distinction, a concept animated by Bourdieu (Arsel & Bean, 2013; Hennion, 2007; Probyn, 2012). For Bourdieu, taste is constructed through myriad factors but driven primarily by one's socio-economic position. Bodies and the tastes they enact come to be representative of the standing of individuals, viewed as markers of social identity and societal position, bound up in, and expressive of, class. As Bourdieu notes:

> [t]astes in food also depend on the idea each class has of the body and of the effects of food on the body, that is, on its strength, health, and beauty... It follows that the body is the most indisputable materialization of class taste.
>
> (1984, p. 190)

While here, I critique this narrow focus and suggest this analysis runs the risk of perpetuating anthropocentricism, this is not to say that taste isn't deeply embedded in sociological relations, nor to deny the significance of power relations inherent in the production of class. These do have significant impacts on the formation of tastes and processes of tasting but a focus solely on structural and discursive renderings tends to obscure conceptions of the materiality of taste and its necessarily embodied mode of practice that move beyond humanist understandings and, as such, fail to recognise the capacity for both bodies and matter (or, more specifically, the relations that enable these to take shape) to act back. Sole concern with structural and discursive manifestations of taste employs a limited understanding of the social (similar to the limitations identified in Nussbaum's capability approach in the preceding chapter) as relating only to particular bounded entities. Through my material-semiotic concerns, I wish to extend the purview of the social to include the relational flows that can enliven entities, rather than solely the bounded manifestations these produce.

My focus on the embodied material relationality of taste is not an attempt to position it as a product of pre-ordained physiological, 'natural' processes. In fact, as I hope I have made clear earlier, the approaches drawn on and extended in this text reject manifestations of such simplistic nature/culture binaries. Yet, we must continue to engage with these tropes because as Teil and Henion note the majority of research on food reproduces a:

> ...nature-culture approach: either food products are just things and their properties are analysed through laboratory tests and measurements; or they are simply signs, the media for various rites and mechanisms of social identity, in which case their physical reality disappears in the analysis.
>
> (Teil & Hennion, 2004, p. 20)

The most significant problem with both of these major stances (if you will forgive me for generalising here) is the construction of taste as relatively stable and able to be clearly traced to specific triggers. Moreover, the focus on the role of individual, bounded entities as actors (bodies or matter) belies the relational becomings of taste. Taste is enacted through relational entanglements of flows—inclusive of the food, the process of eating and digesting—that enable these relations to exhibit forms of vitality. Due to variations in contexts, taste does not have universal nor stable effects or affects on bodies.

In this chapter, I focus attention on taste's capacity to induce affects and how its embodied, relational mode can facilitate an opening up to attunement among human and more-than-humans. This is enabled through visceral and sensorial encounters with relational becomings that, following & Hayes-Conroy, can enable us 'to make a powerful link between the everyday judgments that bodies make (e.g. preferences, cravings) and the ethico-political decision-making that happens in thinking through the consequences of consumption' (2008, p. 462). Indeed, both bodies and the materials we ingest are key players in the becomings of taste and how it comes to matter.

The becomings-with of taste are necessarily relational and multispecies endeavours. Taste brings us into contact with unknown and unknowable others and stretches spatio-temporalities. These 'excessive' more-than characteristics of taste, unable to be contained within singular bounded entities in the here and now, along with the encounters with alterity (our own and others) it enables and the affects these enact, have the potential to support attunement among human and more-than-humans. As Goodman suggests, such manoeuvres could be the basis for new world-makings:

> Because taste and the 'feeling of food' is indeterminate, albeit conditioned and contingent, it opens up spaces of hope for greater understanding, appreciation of difference and acceptance that, ultimately, might ground a progressive politics of change.
>
> (Goodman, 2016, p. 260)

This 'hope' and the potential for more ecologically sensitive forms of living is missing in the rather static readings of taste presented through anthropocentric 'nature-culture' and sociological representations that dominate narratives of taste and food more broadly.

Sensorial, responsive experiences of taste run counter to dominant anthropocentric representations while their very materiality also positions them as somewhat resistant to narrativisation. Yet, if we are to lay the groundwork for new ontologies and world-making practices, we need ways of speaking about these experiences that can support and amplify these alternatives. The challenge is material-semiotic in nature and requires the development of new grammars attuned to materiality and embodiment to represent and support these relational, responsive modes of being and doing. Varying Goodman's assertion ever so slightly, in this section I suggest a good place to start is with a focus on new 'grammars of taste' (substituting taste here for Goodman's eating).

To do this, we first look to the relational entanglements that have induced a 'dulling' of sensitivity to taste through the rise of international agri-businesses in the twentieth century. This process highlights both the contingent and political nature of taste along with the need for more nuanced attention to its affective force, particularly through close attention to taste's viscerality. Here I draw specifically on the work of the Hayes-Conroys' and Michael Carolan before dwelling with Elspeth Probyn's notion of taste as attachment (Probyn, 2012). Following on from the critique in the previous chapter, attachment and associated notions of care will be problematised. An alternative approach is presented that explores taste as forms of togetherness predicated on relational flows that can be realised through playful tinkering enacted with the ethico-political notion of convivial dignity.

The shaping of tastes

It is generally accepted that the economic and, consequently, the social shaping of tastes today have been significantly impacted on by the rise of international agri-business throughout the twentieth century. These processes have greatly reduced the varieties of food commercially available due to an emphasis on economies of scale that require the mono-cultural production of food that can withstand long transport times. In fact, the rise of international agri-business, which Campbell, (2009) identifies as starting with the industrial revolution and reaching their peak in the global food system of the twentieth century, has largely obscured the inputs (the human and nonhuman components that contribute to its becomings) required to produce food. Campbell goes so far as to say that international agri-business has rendered food invisible (2009, p. 313). Throughout the twentieth century, this invisibility aided and abetted the overarching drive for cheaper food, often resulting in production occurring at ever greater distances from places of consumption and prompting an increase in levels of processing, both evident in the growth of the fast-food market in the latter decades of the century.

This so-called invisibility of food is aided by, as well as contributes to, what Carolan, (2015b) calls the 'forgetting' of food. This forgetting includes not only the loss of production of particular varieties but also of their taste due to the loss of the cultural and practical knowledge of how best to cultivate them and how they can be prepared. In these ways, the forgetting of food extends beyond the present. It is rhizomic, reaching back into the past, extending into the future and, in fact, paying little heed to linear time at all. Foods, and potential enactments of their taste, can be 'lost' even when their seeds may be 'saved', locked safely away in airlocked seedbanks in undisclosed locations. Taste is not simply about the material itself, rather it is about matter-in-relation. As Carolan writes:

> As crops and management practices stop being used they risk being forgotten. Conventional gene and seed banks are not saving nearly enough, which is why things like memory banks are starting to be discussed with greater frequency (see e.g., Rhoades and Nazarea, 2006). What good is a seed or a bunch of 1s and 0s once divorced from the socio-cultural webs whence they came? Can you tell by looking at a seed how deep it ought to

> be planted or how it responds to certain climatological events? What about the taste, texture, and mouth feel of the fruits that it will bear? Or how it ought to be harvested? Can you get any of that by looking at a DNA sequence? No.
>
> (Carolan, 2015b, p. 130)

The materiality of food and its taste is reliant on more than its DNA and nutritional make-up. However, encounters with food's visceral qualities—its 'taste, texture, mouth-feel'—and its very 'relational contingencies' that mark out its indeterminate qualities (Goodman, 2016) signal opportunities for being and doing differently, but these have been greatly reduced by industrial agriculture. Drawing on Thompson, Carolan points out a steady decline in the number of varieties of foods grown with most Western diets able to be traced back to around ten plants (Carolan, 2015b, p. 130). The foods most commonly available for those in the minority world to taste have become those that pack and transport well. This is often those able to be picked prior to peak ripeness to conform to the needs of stretched production and distribution networks. For Carolan, this system 'flattens out the tastes and experiences available to consumers' (2015a, p. 321). This very flattening-out of food-based experiences, including the socio-cultural and socio-economic context such as the characteristics of where food is purchased, prepared and consumed, impacts on the enactment of taste by reducing exposure to different foods and the situations in which we can encounter them. As Probyn suggests, 'The Westernization of diets around the world has arguably homogenized tastes and produced a putatively global epidemic of obesity, with effects felt particularly by the poor and often by women' (2012, p. 58).

The homogenisation of tastes has been commonly represented in popular discourse, as a result of an innate human desire for certain foods, namely those that are sweet, salty and laden with fat (see Schlosser, 2001). This 'hardwiring' is then seen to be exploited by fast-food companies and those producing other highly processed foods, leading to consolidation of human 'addictions' to unhealthy food and, as the narrative goes, contributing to a global obesity epidemic. However, taste is identified in much of the literature explored here, from Bourdieu to Carolan, as developed through 'training'. For Bourdieu, social distinction is produced through forms of training which are not necessarily overt but produced through habitus and doxa of particular fields. They are, indeed, learnt. Carolan conceives of tastes as a form of tuning, noting that 'bodies and societies needed to first become ""tuned"" to industrial food for this system of food provisioning to have the grip on us that it currently does' (2015a, p. 318). However, through his attention to the viscerality of taste, he identifies both the potential and need for a retuning. In so doing, Carolan encourages the development of new normative approaches to eating that enable us to 'taste' differently. Such retunings necessitate the opening of alternative possibilities to play with food grounded in embodied, visceral engagements.

Making the visceral matter

The turn to the visceral and its affective forces aims to encourage greater attention to the complexities of food and its relations with bodies and the world more broadly. For me, as Goodman, (2016, p. 258) suggests is also evident in Carolan's work, concern with, or sensitivity and attunement to the visceral is accompanied by, and in some ways indistinguishable from, a focus on matter and its affective force. In this vein, the ideas developed here build on and contribute to a fleshing out of what Goodman refers to as a growing body of work on the '"vital" (re)materialities of food' (2016, p. 258). This vitality of matter is experienced through multi-sensory engagement with its very materiality. Responsive attunement to matter is an embodied experience that, in relation to taste, is marked by complex multi-species body/mind relational entanglements. The visceral is key here and I follow Probyn (as summarised by Waitt) in attempting to 'highlight the visceral to trouble what is knowable' (Waitt, 2014, p. 411). Before we go too much further, let me make clear that I draw directly on the Hayes-Conroys' definition of visceral as:

> . . .the realm of internally-felt sensations, moods and states of being, which are born from sensory engagement with the material world. Note that we include in visceral experience the role of the cognitive mind; visceral refers to a fully minded-body (as used by McWhorter 1999) that is capable of judgment.
>
> (A. Hayes-Conroy & J. Hayes-Conroy, 2008, p. 462)

Bodies do not exist or 'do' in isolation and mind driven actions are always embodied. This is particularly apparent in relation to food. The material and visceral nature of what bodies do with food and foods to bodies, focusing here on taste, stretch out to a vast array of broader concerns. As demonstrated through the methodological framework termed a 'political ecology of the body' developed by & Hayes-Conroy:

> . . .different tastes for certain foods are also materially developed in the body, brain, and tongue, and are articulated through complex assemblages of past opportunities and vacancies, personal memories, social histories, and random events.
>
> (2013, p. 83)

Such processes challenge common assumptions that 'sensory modalities exist as a natural/essential category that is both prior to and distinguishable from their social experiences and intellectual development' (J. Hayes-Conroy & A. Hayes-Conroy, 2013, p. 82).

Through their focus on the visceral, the Hayes-Conroys aim to develop an understanding of how bodily decisions and affects relate to the broader 'rigidities of our socio-political world and yet remains open to the new possibilities that affective encounters may allow' (2013, p. 82). However, moving from openness to action can be challenging. As Mol notes of her enduring dislike for granny smith apples—entities imported from Chile to the Netherlands during the Pinochet years —'Granny Smith apples came to taste of violence' (Mol, 2008, p. 33). She still does not like them, observing that 'it turned out to be difficult to re-educate my taste' (2008, p. 29). The enduring embodied legacy of 'bad taste' can prove as hard to shake as the consumption of things identified as 'tasting good'. Those who have a taste for 'junk foods' or other industrially produced products cannot simply be dismissed as 'not using their senses properly (or have[ing] forgotten how to)' (J. Hayes-Conroy & A. Hayes-Conroy, 2013, p. 82). Still, through attention to the visceral, we can see projects such as the school kitchen gardens the Hayes-Conroys analyse as not only neoliberal actions offering 'better taste' (though many of course do see them in this instrumentalist way), but as offering 'students a chance to have novel experiences with food/practices which allow them to interrupt current habits of bodily (re)action and begin to feel out different ways of being and becoming' (J. Hayes-Conroy & A. Hayes-Conroy, 2013, p. 84).

The affective force of these encounters offer fuel for 'visceral imaginaries' (2013, p. 84). As such, a willingness to enter into risky relations with the unknown, conceptualised here as openness to new tastes and attunement to their affective forces may be able to support alternative forms of understanding of the human subject and engagement with the world. These manoeuvre can be conceived of as forms of playful tinkering marked by openness to being moved by others and the relations through which all entities take shape. As identified in the preceding chapter, engagement in play also offers a mode of encounter that supports multisensorial, responsive training capable of developing the skills and capacities required to be resilient in the face of uncertainty. Play also has the capacity to act as a 'charge' to ignite creative imaginings of the world through careful attunement to the generative potential of relations.

Playful tinkering with attunement, affordances and training for uncertainty

Attunement to the affective force of ongoing embodied, visceral encounters with the materialities of new foods and their tastes requires sensing and adjusting of relational encounters that may be capable of enacting and generating something new. These are all hallmarks of play as a variant of experimentation. As Anderson and Wylie state:

> An attunement [is] how heterogeneous materialities actuate or emerge from within the assembling of multiple, differential, relations and how the properties and/or capacities of materialities thereafter become effects of that assembling.
>
> (2009, p. 320)

Attunement is necessarily productive. Of course, we must be alert to the potential for vastly different power relations among entities and contexts to become apparent, but the process of attunement can respond to these in a variety of ways depending on the affordances and capacities at play.

Like attunement, affordance is about embodied relations and what can happen—or what relations can form and the effects and affects they can have—when certain encounters among humans and more-than-humans unfold. Playing with taste, thus, can induce a variety of affordances that offer possibilities for action, of becomings-with, unknown and unknowable entities and relations, the outcomes of which cannot be predicted. Though certain patterns could become rather stable over time, there is always the capacity for shifts to occur. As Gibson writes:

> An affordance cuts across the dichotomy of subjective–objective and helps us to understand its inadequacy. It is equally a fact of the environment and a fact of behavior. It is both physical and psychical, yet neither. An affordance points both ways, to the environment and to the observer.
>
> (2015, p. 121)

The generative capacity of affordances as a component of play can support the development of alternative, attuned modes of being and doing. There is also the potential here to train bodies to be alert to the possibilities of affordances. This potential is particularly apparent in relation to our necessarily, everyday interactions with food. Indeed, affordances can only be enlivened through embodied experience. They are visceral encounters that relate to practice—to doings—such as eating and tasting.

This 'doing' and the notion of the visceral are necessarily relational rather than reliant on a pure 'natural', 'innate' sensorial response. Indeed, keeping food firmly in our sights, senses present as particularly fruitful sites for training as Probyn writes of taste:

> The body that tastes is also a body in training. It's clear that through learning to taste people also acquire a different sense of their bodies and of the world, as well as of the different ways of describing how they fit within a complex web of things and tastes. In other words, through taste our bodies learn to be affected, and to reflect upon how and what affects us.
>
> (Probyn, 2012, p. 67)

She goes on to note:

> As the body becomes a sensitized medium, relations between different parts of the body vibrate differently in tandem with different aspects of the thing being tasted. This activity remakes the world, makes up new worlds.
>
> (2012, p. 67)

Affordances are visceral experiences that can be attuned to and worked according to skills and capacities (including sensitivity). Our attunement to our necessarily visceral encounters with food through modes of playful tinkering provides opportunities where these skills and capacities to 'remake' and 'make up new worlds' can be developed through training. This training may be about the potential of particular entities but this is not to be understood simply as reliant on the inherent qualities of particular actors.

Instead, the necessary skills and capacities can only be enacted as part of relational flows. In taste, this means they are formed through the possibilities afforded by bodies, materialities and contexts and are open to being worked and reworked. As a consequence:

> ...different bodies have distinct affective capacities. The intensities of such forces can vary as they pass between, or through, and come to inhabit different bodies, and may be narrated in various ways as they are understood within particular contexts—such as care, shame, disgust, pride and love. The visceral realm is therefore about being able to position oneself in relation to others, and things that shape and reshape understandings of food, subjectivities and spaces.
>
> (Waitt, 2014, p. 412)

These capacities are, of course, not static. Attunement and affordances in forms of play enable them to be worked on, to produce sensitised bodies-in-training. The training never ends as the relations are able to constantly shift and move. In this way, we can conceive of the relational experiences of taste as forms of play attuned to uncertainty.

Moving beyond attachments, detachments and care

Taste is relational and these relations stretch across space and time connecting individual bodies in specific locations and periods to others elsewhere in a multiplicity of ways connected to memories, place and myriad other events. Taste does not fit neatly inside bounded individual bodies. Its very sensorial viscerality constantly pushes it out and in this way individual bodies, as Probyn drawing on Deleuze and Guattari contends are rhizomatically produced and connected. These connections, she suggests, are crucial to the formation of taste and tastes, which Probyn reframes as forms of attachment:

> ... to the strangers who grow or produce food, to friendships which are made and remade in commensality and sharing, to the soil and water, the factories, ships, oyster leases and abattoirs. Taste ties us to history just as it does into various economic relationships.
>
> (Probyn, 2012, p. 65)

While Probyn acknowledges that 'taste is a problematic modality of attachment to the world' (Probyn, 2012, p. 65), the focus on attachment as a means to 'move beyond focusing just on humans and instead consider the different connections of human, nonhuman, technology, history, and culture' (Probyn, 2012, p. 66) is also of concern. While many of the arguments developed by Probyn resonate with this text's identification of the need to move beyond the humanist subject, as outlined in the preceding chapter, the notion of attachment troubles me. Modes of detachment and other forms of togetherness might also provide productive ways of thinking through entangled relations and their effects. Connections may, in fact, be more generative in certain contexts when they are marked by a 'holding apart' rather than through attachment which is commonly bound to notions of care and love and an implied sentiment of familiar, knowable intimacy.

Sentiments of love and care may well be able to support political action and productive ontological frameworks (see Gibson-Graham, 1996), but I wonder whether the associated level of 'affection' so readily extended to those like us (as a corollary, so easily denied to those with whom we do not identify or, in the very least, to those we don't find cute and able to be easily loved) is necessary for inducing responsive, sensitive adjustments to others. I worry that attachments, risky or otherwise (Latour), may have a tendency to smooth out alterity in the search for commonalities that can make us sticky together. Probyn riffs off French sociologist Antoine Hennion's work 'to frame taste as "those things that hold us together"' (Probyn, 2012, p. 65), but I am interested in what happens when we also focus attention on the affects and effects of how taste can also hold us apart, not in a sense of hyper-separation, but through a receptivity to alterity that can support other forms of togetherness even when attachment is beyond reach or undesirable. This would require taste to be conceptualised through other modes of doing and being. Yet, a desire for encouraging attachments and connections through modes of care are enduring in work engaging with the food system and everyday food practices.

Care is regularly called upon to explain and encourage connection or reconnection with taste, food and the food system. Carolan points out that '[h]ow we are connected...influences how we care and what we care about' (2015a, p. 318), turning to feminist scholarship and notions of care to 'make sense of affective encounters' (Carolan, 2016, p. 147). However, I remain wary of using care as our driver here for determining what matter matters in relation to taste. While the feminist informed work underpinning the 'ethics of care' represents an important critique of the 'autonomous subject who makes rational choices', as Harbers et al. observe:

> ...so far a humanist orientation has dominated the ethics of care. The relevant entities in its theoretical repertoire are human beings. This implies that while the theoretical notion of the *will* has been subjected to considerable change, its counterpart, *nature*, has been left unanalysed in this tradition.
>
> (Harbers, Mol & Stollmeyer, 2002, p. 218)

It is the enduring humanist connotations of care that continue to trouble me. The affective force of the relational entanglements of taste encountered in my fieldwork are not reliant on motivating humans to care for others. While such sentiments are sometimes evoked, the relations are much more complex and entangled than an ethics of care approach suggests. My wariness of care echoes the concerns raised by Jane Bennett (2010).

For Bennett (2010), efforts to encourage mediation of anthropogentic climate change through practices of care tend to invoke a vulnerable earth that requires humans, in the form of the 'modern' subject, to save it. In this way, the great myth of human exceptionalism that has fuelled destructive engagements with the environment seem to reappear in calls to save it. The notion of stewardship is similarly problematic (see DeFries et al., 2012; Ellis, 2011), demanding the forging of human driven connections. However, there may be other ways to care as Puig de la Bellacasa identifies in her detailed work with soil, 'humans are not the only ones caring *for* the Earth and its beings—we are *in* relations of mutual care' (2010, p. 164). The question, as Puig de la Bellacasa astutely points out, then becomes 'How do we engage with accountable forms of ethico-political caring that respond to alterity without nurturing purist separations between humans and nonhumans?' (2010, p. 159). This question does indeed need to inform the use of conceptions of care, but still my concerns are not allayed.

As I suggest above in relation to attachment, even if hyper-separation is not a feature of care, the concept runs the risk of 'flattening' out alterity by encouraging care for those with whom intimacy, or in the very least familiarity, can be readily enacted. These ideas are imbued with the notion of there being a subject who acts as 'carer'. This is a limitation Mol identifies and one that, linguistically, she has not overcome. Yet her work is, of course, at pains to point out that, despite the notion of care implying the presence of a dominant actor, care is enacted through responsive human and more-than-human relations—people, equipment, spaces, atmospheres. Within these creative, shifting assemblages, particular components, notably (for my purposes here) food and its taste become recognised as being able to provide care. Indeed, food and drink, as Harbers et al. state:

> . . .are media for care—they *do* care. *They* taste good or bad, have a nice or gruesome texture. They are, not as delegates of people, but all by themselves, objects of longing or aversion. It is thus that attending to food and drink in all their daily life complexity is a crucial part of caring.
>
> (Harbers et al., 2002, p. 217)

The relational aspects of food are a little hidden here, but the tasting (and the taste itself), digesting and all of the 'doings' related to food are both produced and embedded within broader webs of relations that enable and enact care. Here we can see that, while care is so often represented through modes of intimate attachment aligned with notions of 'good' and 'bad' care, it can also be enacted through detachment. In the above example, there is a desire to detach from certain foods that lead to particular taste experiences. Taste is produced through responsive relations predicated on modes of togetherness, but these do not require relations of care. Indeed, attunement to relational taste has the capacity to reconfigure notions of the humanist subject and shift interactions with the world to disrupt myths of human capacity for unidirectional, singular action.

However, what is of most significance for my purposes here, and key to the enactment of taste, is not the identification of how particular attachments and detachments might or might not enable realisation of relational care, but the ways in which those within these relations are open to being moved by and responsive to others and the relations that enable these encounters. While some of these encounters could be understood as examples of care, I suggest that this is not the only practice this speaks to. Perhaps a focus restricted to care (particularly through its concern with attachments and desires for intimacy and familiarity) may in fact stymie encounters with being moved through the perpetuation of particular modes of non-reflexive care (of knowing better) or through distancing those who do not identify as being people who would 'care' for certain things in particular contexts, for example, not wishing to ascribe to a 'green' subjectivity or not being able to 'care' for the unseen or the ugly. Training of sensitivities to attune ourselves to the human and more-than-human formations of taste and its enactment through a series of flows involves developing an openness to being moved by humans, more-than-humans and the spatio-temporally ambiguous relations that bring these entities into being and shape their interactions, each responding to the other as a form of unique choreography. Taste involves both movement—the 'hustle of flows'—and a process of being moved by that is not reliant on attachments or detachments but multifarious forms of togetherness. But just how can we talk about these movements?

I suggest the need to develop a broader grammar for supporting and understanding practices of being moved that exceed exultations to care. In so doing, I do not wish to enact an outright rejection of care. The detailed reconceptualisations of this term and practice offered by Mol and her collaborators, along with Puig de la Bellacasa, point to the generative potential of encounters with care. But care so often doesn't incorporate being moved that it needs companions interested in supporting more ethical human/more-than-human engagements. The training of sensitivities necessary for encouraging attunement to taste, and the movements through which it is experienced, can be understood as enacted through forms of playful tinkering. Play, when enacted with convivial dignity, acts as a form of relational encounter that does not require attachments but necessitates responsiveness to being moved and an embrace of in-process state of beings. This exploration aims to initiate a discussion of how understanding taste as generated through play could contribute to the development of the capacities and skills needed to be moved by matter and relations enabling the development of recognition of the necessity of our togetherness-in-relation that could prompt more agile ways of living with the uncertainty and the multiplicity of vulnerabilities this brings to the fore.

Playful experimentations with taste

Play provides an alternative to concerns with attachment and detachment and notions of care as an ethico-political prompt for more generative approaches to living in the world. It is predicated on conceptions of togetherness-in-relation which, when guided by convivial dignity, is choreographed through exposure to mutual vulnerabilities. What we can eat, and the tastes that coalesce, highlight our reliance on multispecies relations, exposing the limits to human control. The growing vulnerabilities of our food supplies wrought by a changing climate serves to further emphasise the contingent nature of what and how we can taste now and in the future. Taste requires us to enter into ever riskier encounters. In response to these, many people around the world are engaging in the development of alternative practices inspired by a taste for different relations for a more hopeful future. These tastes of hope that aim to secure more sustainable modes of living and resist the dulling of tastes induced by 'big food' commonly manifest through forms of playful tinkering with food and tastes that require attunement to their more-than-human and relational configurations in order to prompt reflections on the fragility of human exceptionalism.

Carolan, focusing on agro-food studies, identifies the necessity of engagements in co-experimentations that attends to practice and is unable to be represented through language, making reference to the need to 'get involved in it'—not just know about it (Carolan, 2013, p. 148). Similarly, Atchison develops the notion of 'experiments in co-existence' in her and her collaborators' (including Lesley Head) work with invasive plant species. This approach combines Mol et al.'s notion of embodied tinkering (a form of step-by-step adjusting (Mol, Moser & Pols, 2010, p. 14)) with scientific experimentation, drawing particularly on Stengers. Atchison and Head state that 'Plant bodies challenge and energise human-centred concepts of the body by expressing different forms of collectivity, mobility, and agency' (2013, p. 952). Just as Mol asks, 'what is an apple?', they ask, 'what is a plant?' and encourage the development of 'plant perspectives' that attempt to ignite new formulations of what a body is and how, if at all, it can be contained (Atchison & Head, 2013, pp. 955, 951). As with play, the experimental doings as conceived by Carolan and Atchison et al. are visceral, engaged, responsive and embodied (the latter occurs in a stretched sense as they also aim to reconfigure just what we mean by bodies). If we take up these invitations to reconfigure bodies, then our sensorial conception of taste is also open to creative redoings. In this book, I suggest that playful tinkering, as a variant of experimentation, provides the creative and responsive spark to enable generative modes of being and doing. Taste is a site where this form of play comes to the fore. In my fieldwork, this often centres on shifting modes of food provisioning that respond to seasonal food and experiences of scarcity and abundance. These are encounters for my urban and suburban dwellers that are often ignited by their participation in AFNs.

Building on the Hayes-Conroy's work, the case studies explored in the following chapters do not attempt to glorify or validate alternative food networks as a panacea for current food system and broader environmental crises. They are not offered as ways of 'fixing' taste and reducing obesity. I am keenly aware that there is a great danger that my arguments could be read as promoting a form of 'good food politics' embedded with assumptions that some tastes are better than others, or, perhaps worse still, that consumers with better taste are simply more reflexive and thoughtful about their food decisions (see Guthman, 2003). Guthman's work powerfully explodes such simplistic readings, from her demonstration of the inequities infiltrating the production of 'yuppie chow' salad mix in California to her more recent focus on strawberry farming and fumigants (2017) where commonplace assumptions about exploitative industrial agriculture and the more just alternative food scene are unpicked to point to missed opportunities to promote improved 'lives and livelihoods' (Guthman, 2017).

The playful capacity of the affective force of taste to be generative of more ecologically attuned modes of being and doing through AFN participation is also limited by issues of access. Importantly, access, as the Hayes-Conroy's show, is not simply related to issues of distribution, supply and economic standing, but is also related to embodied visceral encounters and their affects. Indeed, they demonstrate that 'food access in affective/emotional terms is about a whole network or rhizome of forces that influence bodily movement, desire and drive' (J. Hayes-Conroy & A. Hayes-Conroy, 2013, p. 87). In line with this, my analysis is not aimed at making judgements about whose taste is right, more reflexive or socially or environmentally better, even though many of the participants in this research often have plenty to say themselves about these matters. Instead, my aim in this section is to sketch out the detail of the very visceral 'doing' of taste and the 'doings' it induces in those who are food-producing gardeners or AFN shoppers and/or agricultural show exhibitors. What I am concerned with here is the role of relational materiality in these becomings, which, while never able to be divorced from the broader social, political, economic and environmental concerns, also play a role. To work the potentialities of taste, we need to understand the affordances as well as the skills and capacities involved in tasting. I am not offering a new normative notion of taste, but rather I am to encourage a more open conception of taste tied to developing the skills and capacities to both respond to and be affected by others. Taste provides us with an opportunity to 'train our sensitivities' (Mol, 2010, p. 130) to do just this and to encourage recognition of the inescapability of togetherness-in relation.

Conclusion

Attunement to the relational becomings of taste through play can enable the generation of new ways of being and doing that are more than the sum of its parts and exceed the bodies of bounded entities. This may not always be something that is easily palatable or which brings with it an affect of joy. It is certainly not something that necessarily elicits desires to care or to be cared for. We are likely to encounter disgust and revulsion in these becomings and detachments as much as attachments. This is primarily a result of playful tinkering with taste inducing encounters with mutual vulnerabilities where nothing or no one bounded entity is in control. While 'taste' can be learned or 'trained' as a means of attuning oneself to respond to particular taste enactments and specific materialities, this is a training of sensitivities to matter that may not always conform to our previous experiences and which can be decidedly unsettling, particularly for those bound up in myths of human mastery.

To taste is not necessarily to come to 'know' that which is ingested. Taste's alterity, shifting forms and contingent nature muddle any easy understanding here. Thus attuning to taste through playful tinkering provides opportunities for 'training for the unexpected' (Spinka, Newberry & Bekoff, 2001). It invites us to enter into risky but necessary relations with unknown others, the consequences of which cannot be pre-determined. While taste, in a sociological sense, is readily linked to conforming to 'the rules of the game' (Bourdieu, 2003, p. 28), reconfiguring taste as enacted through playful tinkering casts it as propositional, quite often involving a rejection of order and rules to enact something new. Taste itself is shown to be formed through responsive, relational flows. It is through attunement to the relational flows of taste, and recognition of shared vulnerabilities, via playful, tinkering modes of encounter guided by convivial dignity that I sense the potential for the generation of new ontologies and world-making practices.

Convivial dignity is employed to highlight the need to expand the sociality of taste and the contribution made to it by a variety of human, more-than-human and relational components. The unfolding relations are not necessarily (and, in fact, are rarely) marked by equality; and we must indeed remain alert to power, taking seriously concerns raised about the potential for embodied, visceral and affective concerns to flatten out these significant impacts on relations. However, at its heart, convivial dignity is about openness to being moved by others, human and more-than-human. It is about recognition of the necessity of togetherness-in-relation and the development of a grammar to amplify this.

In the gardens and kitchens that dominate my fieldwork experiences and the discussions of taste in the following three chapters, we will see that relations are rarely clear-cut. There are glimpses of both attachments and desires for detachment, but overwhelmingly what I find are efforts to engage in playful interactions with an appreciation of the need to live together regardless of the connections that are made among entities and the relations that produce these. The inescapability of our relational entanglements are brought to the fore and, when guided by convivial dignity, these are shown to be able to contribute to the enactment and amplification of more sustainable modes of living. The affective force of taste is key to its generative potential. To explore these possibilities, we turn first to the stories and practices of urban food producers, followed by AFN shoppers and finally to those who exhibit their home-grown produce in the Agricultural Show competitions.

References

Anderson, B., & Wylie, J. (2009). On geography and materiality. *Environment and Planning A*, *41*(2), 318–335.

Arsel, Z., & Bean, J. (2013). Taste regimes and market-mediated practice. *Journal of Consumer Research*, *39*(5), 899–917.

Atchison, J., & Head, L. (2013). Eradicating bodies in invasive plant management. *Environment and Planning D: Society and Space*, *31*(6), 951–968.

Bennett, J. (2010). *Vibrant matter: A political ecology of things*. Durham, NC: Duke University Press.

Bourdieu, P. (1984). *Distinction: A social critique of the judgement of taste* (R. Nice, Trans.). Cambridge, MA: Harvard University Press.

Bourdieu, P. (2003). *Firing back: Against the tyranny of the market 2* (L. Wacquant, Trans.). London and New York: Verso.

Campbell, H. (2009). Breaking new ground in food regime theory: Corporate environmentalism, ecological feedbacks and the 'food from somewhere' regime? *Agriculture & Human Values*, *26*(4), 309–319.

Carolan, M. (2013). The wild side of agro-food studies: On coexperimentation, politics, change, and hope. *Sociologia Ruralis*, *53*(4), 413–431.

Carolan, M. (2015a). Affective sustainable landscapes and care ecologies: Getting a real feel for alternative food communities. *Sustainability Science*, *10*(2), 317–329.

Carolan, M. (2015b). Re-Wilding food systems: Visceralities, Utopias, Pragmatism, and Practice. In P. Stock, M. Carolan, & C. Rosin (Eds.), *Food Utopias: An invitation to a food dialogue* (pp. 126–139). New York and London: Routledge.

Carolan, M. (2016). Adventurous food futures: Knowing about alternatives is not enough, we need to feel them. *Agriculture and Human Values*, *33*(1), 141–152.

DeFries, R. S., Ellis, E. C., Chapin, F. S., III, Matson, P. A., Turner, B. L., II, Agrawal, A., & Syvitski, J. (2012). Planetary opportunities: A social contract for global change science to contribute to a sustainable future. *BioScience*, *62*(6), 603–606.

Ellis, E. C. (2011). A world of our making. *New Scientist*, *210*(2816), 26–27.

Gibson, J. J. (2015). *The ecological approach to visual perception: Classic edition*. New York and London: Pyschology Press.

Gibson-Graham, J. K. (1996). *The end of capitalism (as we knew it): A feminist critique of political economy*. Oxford: Blackwell.

Goodman, M. K. (2016). Food geographies I: Relational foodscapes and the busy-ness of being more-than-food. *Progress in Human Geography*, *40*(2), 257–266.

Guthman, J. (2003). Fast food/organic food: Reflexive tastes and the making of 'yuppie chow'. *Social & Cultural Geography*, *4*(1), 45–58.

Guthman, J. (2017). Lives versus livelihoods? Deepening the regulatory debates on soil fumigants in California's strawberry industry. *Antipode*, *49*(1), 86–105.

Harbers, H., Mol, A., & Stollmeyer, A. (2002). Food matters: Arguments for an ethnography of daily care. *Theory, Culture & Society*, *19*(5–6), 207–226.

Hayes-Conroy, A., & Hayes-Conroy, J. (2008). Taking back taste: Feminism, food and visceral politics. *Gender, Place and Culture*, *15*(5), 461–473.

Hayes-Conroy, J., & Hayes-Conroy, A. (2013). Veggies and visceralities: A political ecology of food and feeling. *Emotion, Space and Society*, *6*, 81–90.

Hennion, A. (2007). Those things that hold us together: Taste and sociology. *Cultural Sociology*, *1*(1), 97–114.

Mol, A. (2008). I eat an apple. On theorizing subjectivities. *Subjectivity*, *22*(1), 28–37.

Mol, A. (2010). Moderation or satisfaction? Food ethics and food facts. In S. Vandamme, S. van de Vathorst, & I. De Beaufort (Eds.), *Whose weight is it anyway? Essays on ethics and eating*. Leuven: Acco Academic.

Mol, A., Moser, I., & Pols, J. (2010). Care: Putting practice into theory. In A. Mol, I. Moser, & J. Pols (Eds.), *Care in practice: On tinkering in clinics, homes and farms* (pp. 7–26). Bielefeld, Germany: Transcript Verlag.

Probyn, E. (2012). In the interests of taste and place: Economies of attachment. In V. Rosner & G. Pratt (Eds.), *The Global and the intimate: Feminism in our time* (pp. 57–84). New York: Columbia University Press.

Puig de la Bellacasa, M. (2010). Ethical doings in naturecultures. *Ethics, Policy & Environment. A Journal of Philosophy and Geography, 13*(2), 151–169.
Schlosser, E. (2001). *Fast food nation: The dark side of the All-American meal.* Boston, MA: Houghton Mifflin.
Spinka, M., Newberry, R. C., & Bekoff, M. (2001). Mammalian play: Training for the unexpected. *Quarterly Review of Biology, 76*(2), 141–168.
Teil, G., & Hennion, A. (2004). Discovering quality or performing taste? A sociology of the amateur. In M. Harvey, A. McMeekin, & A. Warde (Eds.), *Qualities of food* (pp. 19–37). Manchester and New York: Manchester University Press.
Waitt, G. (2014). Embodied geographies of kangaroo meat. *Social & Cultural Geography, 15*(4), 406–426.

4 Growing a taste for togetherness

Introduction

Drawing on my fieldwork walking and talking with people in their food producing gardens, this chapter identifies the ways that attunement to the relational entanglements of taste can motivate particular food habits and practices. What we encounter in these fertile urban sites is an openness to being moved by taste that is induced by recognition of togetherness-in-relation. This capacity to embrace togetherness with unknown and unknowable others and concern, not only with relations among bounded entities, but with flows that exceed these bodies wherein life is understood to be a series of flows and interdependencies, is shown to encourage the development of ways of being and doing capable of challenging anthropocentric conceptions of hyper-separation. These alternative modes of being are shown to be supported through embodied practices of multisensorial engagement with taste that prompt enactment of playful tinkerings in gardens. These practices are enhanced through the affordances of access to food producing spaces and a stretching of the spatio-temporality of tastes through nostalgia and exposure to a variety of tastes across diverse geographical locations.

Through attunement to embodied, visceral engagements with more-than-humans and the relational flows involved in food production, the affective force of taste is shown to encourage these gardeners to work with their plots and the food it yields in particular ways. For these producers, the production of tasty food is understood to be reliant on collaborative, iterative and ongoing relational interactions. They talk of being responsive to the needs of the soil and the plants, drawing on their sensorial engagement with the nonhumans in the garden to invoke an ethico-political stance that privileges the actant capacities of plants, animals or insects. Yet it is the relational becomings-with and among these bounded entities and their effects and affects that are most evident as they walk and talk me through their gardens. These include expressions of forms of attachments and detachments, or 'holding together' and 'holding apart', both of which, while capable of underpinning ethico-political drives, are shown to be too singular in their approach to capture the messiness of the lived realities in these gardens. Instead, I suggest these gardeners are enacting other forms of productive togetherness capable of generating alternative forms of becoming-with that

DOI: 10.4324/9780429424502-4

disrupt dominant anthropocentric notions of human exceptionalism. While narrativising these in-practice moments is challenging for the participants due to the limitations of humanist discourses, I suggest that these relations can be understood as unfolding through playful tinkering guided by convivial dignity which manifests in practices of making do.

To enliven these arguments, the chapter first looks at the shifting landscape of food in cities, beginning with the gradual distancing of food production from urban centres and leading to the recent rise in urban agriculture. I then seize on the visceral, affective and relational turns to explore how, in my fieldwork sites, a taste for the past and a focus on pleasure in gustatory taste can promote adaptive presents attuned to uncertain futures

The emergence of urban agriculture

There has been a veritable explosion of urban agriculture initiatives in the minority world over the last couple of decades. This is particularly apparent in both the growth of, and attention to, modes of food distribution as evident in urban agriculture's flagship, farmers' markets, as well as through shifting sites of production such as we see in the growth of community gardens coupled with burgeoning interest in food producing backyards and verge gardens. The city of Canberra exemplifies these trends. There are currently no less than 18 public community gardens, nine public housing complex gardens for the exclusive use of residents and 77 food-producing gardens in ACT schools (Turner & Henryks, 2012; Turner & Hope, 2015). Landshare also has a presence in the city, both through its official network as begun by Hugh Fearnly-Wittingstall in the UK as well as in many ad hoc landsharing or 'share cropping' arrangements. Anecdotal evidence suggests that the latter is particularly popular for older residents with backyards in some of Canberra's oldest suburbs that still reflect the original intentions of the city's first subdivision practices where plots were to be large enough for a vegie patch, some chickens and a clothesline. As the upkeep becomes challenging for aging residents, a number of householders are looking for help with their gardens in exchange for sharing the fruits of the joint labour.

Subverting geographies: the distancing of food from cities

The wealth of urban agriculture initiatives represents a backlash to the gradual exclusion of food production from cities that we have witnessed across the minority world throughout the twentieth century (see Steel, 2009). Indeed, a focus on food producing gardening in cities is not a new phenomenon. Allotments in the UK have a long and rich history, 'leisure gardens' in Europe (exemplified by Germany's Schrebergärten) remain common today (Gröning, 1996) and communal food producing plots accompanied initial shifts of people from the country to cities in the USA from the beginning of industrialisation in order to maintain healthy workforces for the burgeoning factories (Lawson, 2005). Dig for Victory campaigns (De Silvey, 2005) in the world wars enlisted the general population in food production to assist the war effort. These were not piecemeal in scale with approximately 40% of the fresh vegetables eaten by American residents in WWII being produced by the estimated 20 million gardens that were established (Lawson, 2005). Some of the growers in my research recall toiling in these gardens during their UK childhoods. However, despite this long history, food producing gardens began to be sidelined in city planning efforts in the latter half of the twentieth century when the narrativisation of the city promoted these sites as bastions of culture best supported by removing people from the need to be involved in their own food production and freeing up land for more commercial uses. Food, or so the logic went, could be grown far away and transported to where it was needed.

There is a symbiotic relationship between the distancing afforded by the industrialisation of food and the growth of cities. Both large-scale food production and cities require large tracts of land and the intensification of populations in urban areas post industrialisation meant more people to feed and the need for the development of efficient infrastructure to distribute food (see Steel, 2009). This, according to Steel, 'emancipated cities from geography' (Steel, 2009, p. 38). The ever-growing distances food travels to reach us enables a severing of links between the foods we consume and the specific climatic and ecological conditions of our urban environs. This has fuelled the simplistic narrativisation of a 'nature/culture' divide and supported understandings of ecological conditions as passive, surmountable, and perhaps most importantly, distinctly separate to human lifeworlds (Blecha & Leitner, 2014; Pincetl, 2007; Steel, 2009). The very stickiness of the nature (nonhuman)/culture (human) divide is made clear by its continued prevalence within the academic disciplines interested in challenging this binary. Head (2012, p. 65), for example, identifies that the 'dominant metaphors—cultural landscapes, social-ecological systems, human impacts, human interaction with the environment, anthropogenic climate change [...] all contain within them a dualistic construction of humans and the nonhuman world (otherwise known as nature)'. The persistence of this language and forms of representation ensures the human/nature divide continues to fuel anthropocentric thinking in relation to cities.

Reengaging with urban food production

However, other forms of engagement and representation are possible. Food scares, growing concern about the environment (Seyfang, 2005, 2006; Seyfang & Haxeltine, 2012), the World Food Crises, desires to support local economies and farmers (Norberg-Hodge, 1991), together with increased attention to health issues associated with modes of production (Goodman & Goodman, 2009), have, over recent decades, encouraged growing citizen interest in the food system and producing one's own. This is also encouraged by growth in media attention to food, particularly prompting an interest in locally sourced foods. While large-scale food production has been gradually pushed further and further away from cities throughout the last century, resulting in the loss of some of the most fertile soils and productive 'food bowls' to housing developments, small-scale growing endeavours continue to rise.

As discussed in the preceding chapter, local food and AFNs are not presented in this text as a panacea for the ills of the industrialised food system and I do not wish to be identified as a purveyor of 'good food' hype, but this chapter does suggest that 'close' engagement in food production where those involved are moved by the unfolding relations through playful interactions has the potential to support, and be generative of, ways of being and doing attuned to togetherness while being marked by encounters with vulnerability and alterity that challenge dominant Anthropocentric narratives and practices. Yet, as Carolan states, involvement in food production:

> . . .affects people differently. Doing is not destiny. Growing your own food no more guarantees that you will start caring more about, say, pesticide exposures among field laborers than being-with animals guarantees that you will shun eating meat (Carolan, 2011a). That realization, however, does not discredit anything just said. A replicable or even predictive politics subtracts when what we ought to be doing is adding to the world.
>
> (Carolan, 2016, p. 148)

I take this drive to add to the world seriously and see in gardeners engaged in this research an attunement to this world that, while difficult to narrativise, already challenges dominant anthropocentric practice in its doings. As Carolan points out, this may not be the case for all growers and I am not advocating that we all need to become gardeners. I am, however, suggesting that these can be sites of engagement with embodied, visceral, more-than-human vitality that encourage practices of togetherness marked by recognition of mutual vulnerabilities. These forms of togetherness, marked by shared experiences of being moved in playful tinkering guided by convivial dignity, may well lay the foundations for the development of the capacities and skills necessary for living in times of uncertainty.

Tastes of togetherness in lively relations

As part of what are variously referred to as visceral, relational and affective turns in recent years, we have seen research (much of it set against the backdrop of the Anthropocene while also engaging critically with this term—see, for example, Head (2014)) on the embodied, 'feeling' or 'sensorial' experiences of human and nonhuman relational entanglements (Bennett, 2010; Gibson-Graham, 2008, 2011; Head, 2012; Whatmore, 2002, 2013). Attention, or attunement, to animals has been key to laying the foundational work here, notably Haraway's work in *When Species Meet* (2008), and also the establishment of the discipline of animal geographies (Buller, 2013a, 2013b, 2016). Much less attention in geography, political ecology and other congruous fields has been directed towards less obviously 'lively' nonhumans such as food, soil and plants. As identified in earlier chapters, there are, of course notable exceptions here, particularly the work of Maria Puig de la Bellacasa, (2017) that explores the links between conceptualisation of soil and modes of care and the development of notions of plantiness by Head, Atchison and their collaborators (Atchison, 2015; Atchison & Head, 2013; Cloke & Jones, 2001; Head, 2012; Head, Atchison, & Phillips, 2015; Hitchings, 2007). Head, Atchison and Phillips reconfigure notions of subjectivity to include plants noting that 'focusing on plant worlds shows how human-centred our conceptualisations of agency and subjectivity have been' (2015, p. 404), pointing out that:

> [p]lants challenge thinking about agency and subjectivity against a human norm; in contrast to many animals, plants are so different from us that we are not at risk of confusion. The point is not that plants possess agency, but that they enact distinctive agencies—sun eating, mobile, communicative and flexibly collective.
>
> (Head et al., 2015, p. 410)

The alterity of plants as represented through 'plantiness' suggests that there are indeed other ways of being and doing in the world to which we should be attentive. Enacting this form of attentiveness could contribute to a rethinking of anthropocentrism and responsive adjustments to the way humans 'be' and 'do' in the world. As Atchison et al. write:

> Plants have the potential to energise our thinking about new ways of living in the world, but this will require increased recognition of the planty subjects with whom we cohabit, as well as greater ethical engagement with questions of our mutual living and dying.
>
> (Atchison & Head, 2013, p. 965)

Such attentiveness, however, is not easily cultivated and enacted due to the inability to capture planty accounts of themselves. Though some have tried: Hartigan attempts to interview plants, Tsing writes as if she were fungi and Pitt uses automatic cameras to see what plants do when humans aren't there. Head et al., (2015) see their task as following plants, with a focus on weeds, to explore human-plant interactions.

Human/food encounters in the garden

In the research discussed in the remainder of this chapter, I attempt to explore human-plant relations by following the humans as they show and tell me around their gardening plots. In so doing, I am interested in the ways these gardeners conceptualise and articulate the 'doings' of plants and their becomings-with them. While their responsive, playful interactions with the more-than-human elements of the garden indicate a sensitivity and attunement to relational 'planty agencies' (Brice, 2014, p. 944) and their capacity to 'act back' (Head et al., 2015), these growers struggle to conceptualise and represent this in our discussions even when it is a significant feature of the garden tours ('smell this', 'see what has happened here', 'look what that's done in response to me doing such and such'). These struggles speak to the narrative limitations induced by dominant Anthropocentric framings and point to the need for more expansive grammars, one version of which is represented by the concept of convivial dignity developed in this text.

To tease out these issues, I explore the visceral engagements with food and taste through ethnographic research gathered from in-depth, semi-structured interviews and garden tours with 31 growers in urban agriculture initiatives in Canberra. The capacity for food producing plants and the food and taste they generate to 'act' is identified as being particularly powerful in regard to the gustatory sensation of taste—this is highlighted in experiences of pleasure and desires to encounter new tastes as well as when access to these 'tastes' is thwarted by weather, climate or pests.

Despite evidence that people grow or buy food based on gustatory taste, this has received less overt attention as a motivator for food provisioning practices in existing literature (Hugner, McDonagh, Prothero, Scultz & Stanton, 2007). Where it is examined, taste is generally seen as a social/cultural phenomenon shaped by the ideas related to the environmental, economic and health concerns mentioned above. However, when participants in this research discuss taste they also refer to notions of freshness, the varieties of food that are available and nostalgia for the 'way food used to' taste. Through such practices, these small-scale food gardeners become attuned and responsive to the relational entanglements that produce food and its taste. In so doing, they exhibit and enact a taste for togetherness-in-relation facilitated through playful tinkering and embedded within convivial dignity's recognition that survival is induced by and depends on these relational interactions.

Being moved by food

The small-scale urban producers encountered in this research are moved by food. The visceral and sensorial experiences of taste are key to encouraging this openness and responsiveness with the affective force of food being evident in all conversations with gardeners. Quite often, as I was roaming through gardens with the participants looking, smelling, inevitably tasting and chatting, at the end, usually as I go to turn off the recorder, they remark that they have told me their whole life story. Sharing their gardens prompts connections to their broader life experiences, often involving recollections of youthful encounters with food. These are regularly, but by no means always, remembrances of their parents growing and preparing delicious food. For some there are memories of 'tasteless' food marked by meat and three veg. Common to all, however, is the ability (often articulated for the first time in these interviews) to pinpoint when food became something 'more' to them.

Excitement and responsive approaches to food are tied directly to access to eye-opening tastes among these growers. This was regularly linked to a history that enabled them to train their sensitivities to be alive to food and stretched the spatio-temporal enactment of taste through the recounting of memories and expressions of nostalgia. In the course of our conversations most growers could identify key exposures to tastes that had encouraged them to become more attuned to food throughout their lives. This was frequently linked to stories of travel when they were younger or when traditional Anglo meals were interrupted by new taste sensations, as was the case for Jenna:

> I grew up on a very traditional Scottish diet of boiled mince, which is the worst thing I've had, and potatoes and any other vegetables kind of cooked to death. . .So nothing with any taste. And then, when I was about 14, my mother had a Sri Lankan friend and I went to their place for a Sunday lunch and I can remember vividly seeing this table, like it was a big family and aunts and uncles and this big table just full of food. And I was blown away because there were colours in the food, like the food had colour and texture and life. And the most vivid experience was picking off the plate what I thought was a bean, a green bean, and bringing it to my mouth and seeing everyone go no, it was actually a chilli. That was my first experience with a chilli. And from that moment I was just hooked and obsessed with finding these different foods. . .

Those who grew up in families that had always been involved in some level of food production expressed an embodied attunement to food that continued to play out in their contemporary gardening practices. They could not remember a time when seasonal rhythms and the resulting tastes of foods were not part of their lives. However, this was not seen to be a form of static knowledge and experience. Engaging with food production involved active and changing encounters that required responsive interactions. Tastes of foods differed year to year and within seasons themselves depending on rainfall, temperatures and the form of human labour invested. Being engaged and responsive to food was not seen to be the same as knowing food and was far removed from notions of 'control'. Food producing gardening was, in fact, constantly identified as being marked by surprise and something entered into in a playful manner. These gardeners learnt to expect the unexpected and, while this often induced frustration, it was also critical to what made gardening a pleasurable experience. Navigating and accepting this uncertainty required deep, embodied engagement and sensitivity to perceived plant needs. This does not mean that the garden was viewed simply as a site of 'nature gone wild'. Acceptance of a lack of ultimate control is not the same as simply letting things 'go' (though this did also sometimes occur). More commonly, the garden was a site of adaptive gardening habits developed in response to the plants and marked by a form of playful tinkering.

Adjustments in food production

Following Mol (2010), tinkering here involves bit-by-bit adjusting that occurs through an ongoing process of 'trial and error'. These playful tinkerings include efforts to adjust soil nutrient levels such as nitrogen, modes and frequency of watering, development and implementation of bug traps and garden structures and experimentation with different varieties and times of planting. Taste is a key indicator of the need for such adjustments (though often coming late in the productive stages of the plants, the adjustments are often made in subsequent seasons). As I was escorted on a tour of the community garden plot of Dinah, a mother of four children and the garden assistant at a local school kitchen garden, it quickly became evident that tasting was, for her, an integral component of gardening and a way of sharing her garden and gardening knowledge. While we also looked, touched and smelt, she constantly passed things to me to taste or urged me to pick something and pop it in my mouth. As she handed me some lime basil (something I had never before tasted), she lamented that it wasn't at its peak, noting 'It tastes better in the morning' before the sun concentrates its oils. It was most often the taste, not the look or smell, that continually caused her to stop and think about what the plant 'needed' or when it should be harvested or, more precisely, where she needed to tinker. Though not in these terms, Dinah spoke of taste as relationally produced, expressing recognition that the impact of her adjustments was limited by the more-than-human entities with the final outcome being relationally produced by the interplay of input and responses. To demonstrate these ideas, I recount part of our tour chat as she twisted off a head of corn and peeled back its husk for me to try. The corn had not done what she had expected:

DINAH: And the other two rows, they're just sporadic this year.
INTERVIEWER: Okay.
DINAH: Yeah. So I don't know. I put lots of moo poo down. We had some. . . a bit of dry weather.
INTERVIEWER: Right.
DINAH: It's not as sweet as I'd like it either.
INTERVIEWER: Okay. Do you think that is a water thing or. . .?
DINAH: I think it is.
INTERVIEWER: Yeah.
DINAH: It just needs a bit more water.
INTERVIEWER: And what about mulch, what are you putting down as mulch? You've got. . . thank you [The respondent handed the interviewer some corn to eat].
DINAH: It's just straw, but I'm gonna try and get some third or fourth cut Lucerne.
INTERVIEWER: Okay.
DINAH: It's got more nutrients in it and the soil. . . I think the straw is at least drawing nutrients—nitrogen mostly. It's a bit ordinary. There's not a lot of taste there.
INTERVIEWER: Well. . .
DINAH: I think that's got something to do with the soil.
INTERVIEWER: Yeah. It's not as. . . no, look, I mean it looked like it was juicier than it tasted, doesn't it?
DINAH: Hmm, I know.
INTERVIEWER: And it isn't very sweet. Isn't that interesting?
DINAH: I know. So, I. . .

INTERVIEWER: It's still a nice flavour though. Yeah, that's right.
DINAH: I'll hunt through and see whose tastes nice.
INTERVIEWER: Hmm.
DINAH: It could be the variety. I think I'd already had the seed for a year or two.
INTERVIEWER: Okay.
RESPONDENT: It could be that as well.
INTERVIEWER: Yeah.
DINAH: A question of the reliability of the seed.

The taste of the corn surprised Dinah. This prompted reflection on water, soil nutrients (and how to attend to these) as well as the actual seeds themselves that had been proven to produce great tasting results in the past, but not this year. In response to the taste she was encouraged to taste more, to seek out the best tasting in the garden (this was to involve raiding—with permission—other plot holders' gardens) and investigate their practices to assist adjust hers. The corn looked good. It was plump and juicy when the husks were peeled back and was growing on strong, healthy plants. The heads were quite abundant indicating effective pollination but the kernels were chewy and bland. The taste prompted Dinah to expand her tasting, gather information and formulate a plan for responsive adjustments to improve it. It was to be a process of playful experimentation marked by tinkering.

Playful tinkering with taste in responsive gardening

Relational, adaptive tinkering and adjusting are forms of experimentation that mark many food-producing practices at multiple scales. Both tinkering and play question dominant representations of food production as human mastery and thus encounter challenges in the discursive renderings of these materially attuned alternative doings. Counteracting dominant Anthropocentric practices and narratives present as material-semiotic challenges that require action on both fronts. While Mol sees tinkering as a component of care, as hinted at in the previous chapter, I suggest invocation of care as a modality of relationality can inhibit disturbances to anthropocentric normativities. Instead, I attempt to narrativise this differently through the concept of play. Tinkering can contribute to this notion of experimental play which is marked by a particular form of open-endedness, a desire to see 'what would happen' if I did this or that. However, play can also be less immediately about its material manifestations than tinkering and also aims to work against the identification of a key actor (i.e. the 'tinkerer') or the intentions of identifiable bounded entities. Play is more concerned with the flows among entities and the generative capacity of these responsive interactions. In these ways, both the relational entanglements of the actions and process—how will the seed respond? How will the weather impact on this? How will the frequency of watering effect the plant? What will the affects and effects of our relational encounters be?—and the lack of a specific goal are emphasised. The focus is distinctly on what might happen.

This generative potential of playful tinkering is particularly apparent through engagement with mutual vulnerabilities in gardens. There is a riskiness to these encounters where death, disease and poor production are constant threats. Most notably, however, risk for gardeners is evident in a letting go of myths of human control and hyper-separation. Simultaneously, the gardens also provide a 'safe space' for these risks to be taken. None of the growers in this research are solely reliant on their plots to feed their families, though many do aim to eat mainly from them and talk of self-sufficiency goals. As one participant noted, if a crop fails to produce, or to produce the right taste, 'there is always something else growing'. These gardening practices are marked by playing with taste in a tinkering fashion.

Growers consistently speak of the taste of their own-grown produce in relation to that sourced elsewhere and assumed to be produced through mainstream broad-scale farming. As one backyard gardener and successful Royal Canberra Show exhibitor notes: '[e]verything that you put [grow] in the garden [has a] better taste than from the market or from the shop'. The extent of this difference was often a surprise for the gardeners: 'I never knew a home-grown potato could taste so different from a shop-bought potato until I grew [my own]... and I couldn't believe the taste'. Food sourced elsewhere (most often identified as supermarkets in this research) was identified as being flavourless. As one participant bemoaned, potatoes and strawberries bought from supermarkets taste the same as one another (Aada). Growers directly linked this to a lack of human responsiveness to plants, seen to be a necessary (though undesirable) outcome of the spatio-temporal characteristics of industrial agriculture. This higher 'eyes to acreage' mode of production was identified as necessitating the use of pesticides and herbicides that were repeatedly identified as detrimental to taste. One backyard grower and show exhibitor states that in his food 'There's better taste... because they haven't got the chemicals in them, not much spray, not much fertiliser, for that is better'.

While not all growers followed organic methods and most did not emphasise this as a key driver of their gardening practices, they can all be described as low artificial in-put gardeners. They engage in an embodied, responsive form of gardening with regular, detailed attention paid to each plant. This was described to me as being part of their daily routines and, as I walked their gardens with them, these practices were always on display. As growers directed my attention to particular plants, leaves were twisted and turned often resulting in the squashing or removal of bugs, fingers were sunk into soil prompting adjustments to watering and protective barriers such as shade cloth were erected. As one community gardener summed it up '... we prefer to...garden in a way that naturally strengthens the plant immune system'. For him, this required working with soil microbes, prompting him and his partner to invest significant bodily energy into assisting production of high quality compost akin to conducting what he referred to as 'homeopathic' gardening. Through a responsive approach to the 'needs' of plants, the soil and other nonhuman elements, the plants were then perceived to deliver 'vitamins and minerals' to the gardeners, packaged in tasty food.

The responsive, spatio-temporal specificity of playful tinkering with taste means that these practices manifest in a myriad of ever changing ways, but some recurring themes were evident in the gardens. This included nostalgia for the way food used to taste in their childhood; a desire to broaden tastes by seeking out heirloom varieties and reconfiguration of their daily schedule and bodily habits to accommodate food production, preparation and preservation. These practices cannot simply be conceived of as political acts driven by a desire to resist modern consumerist society. Nor as a nostalgic desire to return to 'better' times. While elements of these may inform these practices, the gardeners' responsiveness to taste was also attuned to contemporary experiences and uncertain futures.

Tasting the past, adaptive presents and uncertain futures

Stories of gardening were often represented through life stories punctuated by references to specific locations and the people and food experiences that populated these. While the study of memories constitutes a field of studies unto itself, detailed inclusion of this is beyond the scope of this chapter. Instead, I am interested in how these past experiences are enlivened through contemporary bodies showing and telling me around their gardens with a particular focus on how these draw on, and relate to, conceptions of taste. Tastes of childhood were commonly remarked upon from stories of being introduced to a rich diversity of produce by market gardening grandparents in Italy through to memories of growing up in households where food was little more than fuel and regularly bland and unappetising. This stretching of the spatio-temporal boundaries of taste through the invocation of memories and, at times, nostalgia prompts these producers to engage in playful practices of tinkering in their gardens in the present with an eye to the future. The forms of togetherness encountered in these food-producing sites are marked by practices of humans and nonhumans being mutually moved by each other through the formations and reconfigurations of 'tastes'. While these relational flows elicited through play invoke attachments and detachments, the connections are not limited to bounded entities. Instead, these responsive entanglements represent a broader approach that brings mutual vulnerabilities to the fore and recognition of the necessity of togetherness-in-relation. These modes of interaction and recognition are represented here through the notion of convivial dignity.

This theme of nostalgia, and the spatio-temporal stretching of taste relations, was evident from the very first interview I conducted in one of Canberra's community gardens located along a key arterial road surrounded by nature park. There are no houses within sight and anecdotally I have learned that many Canberrans think this site is in fact a market garden. The signage is small and unless you pull off the road (not an easy feat) to inspect it you would be none the wiser. On a coolish morning, I arrived at Aada's plot to talk to her, and I use her stories and practices now to flesh out points of commonality identified across all of the growers I subsequently interviewed. At the time of our meeting, Aada was an active retiree who shared the garden with a former colleague and ongoing friend. This had led to a few disasters including the plot sharer recently 'weeding out' Aada's newly sprouted seeds. Playful tinkering in these spaces does not always lead to joyful outcomes. Aada had long wanted to join the garden but family circumstances (her husband had suffered an injury leaving her to financially provide and care for the family, including, at the time, two teenage sons) had prevented her doing so for a number of years. As our conversation turned more specifically to what she was growing, Aada immediately and without prompting linked her current crop to the 'taste' of her Finnish childhood telling me, 'It goes so far back from my childhood, this garden'.

It soon emerged that gustatory taste with a focus on the pleasure induced by specific foods was the main driver of Aada's playful gardening practices. As she told me, 'When you get your products here then they are so tasty. I think I prefer more it is tasty than one that is healthy for me'. Tasty food was directly linked to her Finnish heritage, though she had arrived in Australia with her husband over 45 years earlier on New Year's Eve 1966. The foods she grew evoked links to the tastes of her country of birth and the people who had then populated her life. As we walked around her plot, she pointed out what she was growing, reaching down regularly to squash a slater or to better show me the produce, always relating it back to her childhood:

> ... currants, potatoes, raspberries, yes and lettuce I grew already when I was... in Finland my backyard, but they didn't grow very well and I am so lucky they just grow here. And then garlic I have started to grow here. Yes, in Finland we use a lot of onions and I grow onions as well. It looks that I grow more of those [than] what I have grown in Finland.

The foods themselves, and the embodied practice of gardening, constantly moved her between then and now, here and there. Tastes were held in tension between these spatio-temporal rememberings and contemporary practices, between excitement at the more-than-human differences in climate that enabled some foods to flourish in her Canberra garden and frustrations that plans to reacquaint herself with past tastes were thwarted by different geographies and pests. The seeds of her current food-growing practice in the garden were sown in the past but adapted to her contemporary circumstances. As she observes when reflecting on the role of current community gardens:

> And this is lovely to come here because you get new ideas from the other people and then elderly people they are alone at home, many people they are alone at home and if they've got this one where to come and see the people I think that this is really good. When I was a child and we went to visit my aunty, 200 kilometres away from that town where we were living, in that town there were community gardens already. It was 1945. I think they should put all those schoolyards they don't use any more schools [A number of schools had recently been closed by the local government] they should put in community gardens.

This nostalgia that evokes a desire for tasting the past is evident in her focus on producing potatoes, redcurrants, raspberries and blueberries (the latter she recalls growing wild in the forests when she was growing up), yet the garden for her was also a place to play—to get 'new ideas' which enabled her to seek out 'new tastes'. Before joining the garden, she had never seen fresh artichokes or asparagus but now grew both in her plot. The day before we met, she had harvested numerous spears of asparagus but as we inspected the bed she noticed a few spears that had been missed and quickly cut one off, handing it to me to eat. Many things in a garden are tasted raw even when, conventionally, they would be cooked prior to consumption (one community gardener noted that her primary-school aged son only ever ate raw corn). As I crunched through the tender asparagus spear, I exclaimed 'Oh, isn't it beautiful?' and it was, in texture, freshness and the vibrancy of the taste. She nodded her head and said 'Yes, there is that taste difference' and as she handed me another 'This [the taste] is different when you grow your own vegetables'.

While Aada's gardening is enacted with a playful tinkering approach, this is rooted in embodied, visceral childhood encounters with food. She shared with me fond memories of working a plot in the forest with her father when she was a young girl, speaking specifically about the potatoes they grew. Potatoes were also the first thing she showed me in her community garden plot. Aada then vividly recalled the heaving redcurrant bushes her family grew later after the death of her father, remarking on how her mother would pick them and how she, as a girl of 12, would then cycle down to the market to sell them. Despite the vital role of gustatorial taste, for Aada, her taste for these foods pushed at the boundaries of her sensorial engagement with and pleasure in the taste of the produce alone. The taste of these foods was intimately bound up in the broader encounters that led her to ingesting these foods. Taste was produced here, out of a multitude of human and more-than-human, present and past relations. She was attuned to these relations and the togetherness of these that induced her 'tastes' for certain foods.

This togetherness across time periods, scales and species-divides can be conceived of as an enactment of convivial dignity. Aada's playful experimentation in the community garden plot was a means of engaging anew with the myriad elements that enable food production so she could make a version of these tastes still available for her and some immediate friends and family. Indeed, this nostalgia was held in counterpoint with her adaptability and flexibility to tastes related to what could be grown in her geographical location under current ecological and climatic conditions as well as being based on what she could learn from others. One member of this garden, a (mostly) retired gentleman of Italian origin, was a prolific producer of artichokes. He had prepared a 'how-to-guide' on preparing them and attached it to the wall in the communal shed. Aada had responded to this opportunity to push her tastes, something she reflected on as she told me about how the Finnish cookbooks she used referred only to canned artichokes, even making mention of how expensive they are. She had to look elsewhere for ideas on how to use her fresh ones and her artichoke plants, despite how much space they took up, had become a feature of her plot.

Playing with the pleasures of taste

While supermarket food was regularly identified as lacking taste by the growers, the industrial agricultural system that supplies it was also commonly viewed as responsible for limiting tastes through the supply of very few food varieties selected, not due to their capacity to induce gustatory pleasures, but for their capacity to be transported over long distances without damage to their physical form and thus saleability. I want to tread carefully here because, while some participants described their gardening practices as giving the 'finger' (as expressed by Harvey) to mainstream agricultural practices and the supermarkets they supply, the motivation driving the gardening of the majority of growers was not self-identified as a result of an overtly political or activist position. In fact, some gardeners wanted to make sure that (even when they were expressing views that could be considered quite passionately political) they garden for pleasure, not for political reasons or due to any specific oppositional identifications. This manifested in people regularly emphasising that they were not 'hippies', 'extremists' or 'activists'. This is not to say that they saw this drive for pleasure (always linked to the taste of the produce) as necessarily existing outside of these broader frameworks (a few participants keenly linked their gardening to revolutionary and subversive practices), but in our in-garden encounters they tended to focus on the embodied, visceral and sensorial engagements and their affective force with food and food-producing gardening. Pleasure was a key driver here and I suggest that this can be understood as inducing a taste for play.

A taste for play and interest in playing with taste encouraged experimental growing, particularly apparent in trying new types of foods (such as the artichokes and asparagus mentioned above) and in the sheer range of varieties grown. This was exemplified by Warren, a stay-at-home father and community gardener, when he recounted his tomato growing escapades in the previous season (that yielded 65 kilos across the growing period):

> I went very quirky and grew lots of... I had some purpley ones and some yellowy ones and some orangey ones and hardly any one's been red. So, I say to the kids what ones would you like to try? 'Do you want to try a red tomato?' If you've got one that's orangey red. And then we made these various tomatoey sauces that, ones that were orangey made a very orangey coloured tomato sauce. The ones that were yellow made an astonishing small child puke-coloured sort of... so, you make this tomato sauce and you're expecting some redness to come through and you get some insipidy, yellowy sort of looking mince and you think it really doesn't look a most appealing colour for a bolognese. So, this year I'm going back to a bit more red action. A bit less yellow.

Playful experimentations do not always lead to positive results, though it was the visual appeal rather than the specific taste of these less common tomatoes that was the issue for this family. The colour had a significant impact on the desire to taste the food (possibly even impacting on the perceptions of the taste). Interestingly, this was only a problem with the tomatoes that were being preserved for later use. Due to the size of the yield, a significant quantity of them were cooked and bottled for use in later cooking. It was not an option for this grower to let them go to waste even when their cooked forms were unappealing. This necessitated devoting a considerable amount of time and effort to trialling different recipes and a shift in approach for the following year when more conventionally coloured tomatoes would be grown. As was typical of all the growers, an unsuccessful experiment did not dampen Warren's enthusiasm for a playful, tinkering approach marked by trial and error:

> I think the whole gardening thing is a thing to be independent and to be different to the mainstream, I guess. To be able to do, to grow what you want to grow. Varieties that you want to grow and I guess we're fairly borderline extreme on our alternativeness in that respect.

Growing what he wanted included heirloom varieties and other foods that were not commonly available requiring seed to be sourced through specialist retailers. This was driven by a desire to expand the tastes available to his family and friends. This was not about simply 'playing it safe' to provide what he knew would be eaten, but about constantly engaging with things he didn't know and working in response to what they did:

WARREN: . . . I thought it was fun to grow things that you don't know about, so I grew some cima di rapa and cavalo nero and had those blanched and then had them with breadcrumbs. . .
The English translation is turnip top, so basically it's like a canola type of plant that you eat the leaf off and so the leaf, like cavalo nero is like cabbagey-like, but the cima di rapa is more spinach-like and so we hadn't really gotten into it the other year. [This year] [w]e thought 'my god this is fabulous'.

INTERVIEWER: Sounds delicious.

WARREN: Yeah, so, it's very tasty. So it [community gardening] gives you that chance to do that and the gardeners from different cultures, you get some idea of what to cook and stuff, yes. I remember when I had the cimi di rapa growing at Cook, the first year I was there, Marco, with his great garden. He said 'You're the first non-Italian person I've seen growing that'. I thought I was pretty cool. I got it from online and thought it was hardly some great personal achievement.

Access to these unusual foods is facilitated by time and drive, as well as having the community garden plot. Warren was chuffed to be growing something that wasn't seen to fit his cultural profile, but as he says, he simply ordered it online and gave it a go.

Expanding gustatory tastes, making do and encountering the unknown

An interest in exploring and experimenting with different tastes and flavours was common across the gardeners. Another gardener, Helen, who sourced heirloom seeds and had a highly productive garden, observed that efforts to source unusual foods had expanded her diet:

HELEN: I've never eaten broad beans until I came down here.
INTERVIEWER: Oh, okay.
HELEN: And I thought well, stuff it. If I've spent six months growing 'em I may as well eat them.
INTERVIEWER: And how did you find them?
HELEN: Oh, they're all right. I must admit you get a bit sick of broad beans. So the first year I was here, I mean I was religious about my broad beans. I kept them all. Last year I went, oh, stuff it. I'm over broad beans now.
INTERVIEWER: Right.
HELEN: I'll. . . but then I decided, well, you can dry them. Why not? So I just pulled them out and I just dried them. I'll just water now.
INTERVIEWER: Yeah.
HELEN: Because I put in a lot of peas and broad beans. I don't know how that'll work.

Here we can see the tensions in claims that producing one's own food necessarily leads to changing tastes to accommodate what was grown and available. While this did occur initially for Helen, she soon became sick of the food. But, instead of simply discarding them, she found other ways of managing the produce. This must have been reasonably successful because being 'a bit sick' of broad beans in the previous season had not stopped her planting them again. However, as she showed me her 'windbreak' of Jerusalem artichokes ('fartichokes' as she called them) it soon transpired that these were a vegetable she had never eaten. They had grown unintentionally; a small segment from a previous gardener had hitched a ride as she was moving soil around her plots. While she described their capacity to spread as 'being worse than cancer', she did not actively attempt to remove them from her garden. Instead, she used them to protect the rest of her plantings from strong winds and set about finding someone (a friend of her mother's was soon identified) who would eat the harvest. So, as we see here, while the expansion of gustatory tastes may not go hand-in-hand with growing one's own, engagement in food production tends to prompt a desire to make sure the produce is used. This speaks to an enacting of the ethico-political notion of convivial dignity where mutual vulnerabilities are exposed through the inescapability of our togetherness-in-relation. The fact these plants managed to grow from what she believes was a 'thin sliver' into a vigorous windbreak prompted the grower to leave them be and seek out someone to whom the produce could go.

In the above examples of the tomatoes, broad beans and Jerusalem artichokes, we see evidence of making do and adapting to what is available and what grows well in response to particular conditions and the specific forms of human and non-human inputs invested in these sites. The gardens and growers are moved by food, and these encounters enable a training of sensitivities that encourage the development of the capacities and skills to be flexible and adaptive to the human and more-than humans encountered in the garden. This was also exemplified by Peter, whose taste for Mexican foods prompted him to seek out certain plants not readily available in Australia. But he did not always have success with his growing as he notes, 'It's a two-way street. I try to grow what I'd like to cook but sometimes it doesn't work out so I cook what we grow'.

This idea of making do was also evident in the sharing of produce, seedlings and seeds among friends and family, acquaintances and within community gardens. Dinah, one of the community gardeners, actually carried out most of her seed swapping via mail through her active membership and connections in an online animal forum. Canberra Organic Growers Society, who oversee the gardens where this research took place, also have an active backyard gardening group and seed saving network of which Dinah is also a member. These interactions provide a means of sourcing foods that as one gardener noted, 'you don't find in punnets at the shops'. As we saw earlier with Aada, Warren and Helen, this sourcing of seeds is not always a result of gardeners 'knowing' a particular taste, but a process of playful tinkering to see how it grows, what can be done with it and, of course, central to all of these is how it tastes. In these ways, there is as a sense of the growers 'becoming-with' the human and more-than-human elements of the garden.

While many of the previous examples tend to suggest a controlled approach to what is planted in gardens (even if what is produced is beyond control), this playful experimentation is perhaps most apparent when gardeners plant things without knowing much about them, sometimes not even knowing what they are called. This occurs frequently in the community gardens where people have extra seedlings to share around. Dinah had planted a number of tomato seedlings grown by her father-in-law, not knowing what type they were. Even Greta, with a highly productive plot in a large community garden on the outskirts of Canberra (who can be described as a very careful, considered and orderly gardener) made space for planting 'unknown' shared seeds and seedlings even when she suspected they were foods she does not have a 'taste' for. As she recounted:

> I don't like Asian food very much. So it's mostly European-type foods that I will grow. . .I don't grow much. . . I've got some Asian greens from this person here [demonstrating] who's new. She's only been here a few months. She grows. . . she's Burmese. And she has about six words of English. And she's been here in Australia for three years. But she gave me a handful of seeds, because every, you know, I give her food if I've got some. I'll give a few carrots or give her this, that and the other. So she gave me a handful of seeds. And when I asked what they were, and she had her niece with her that does speak English

> at the time, and I asked what they were, but the niece didn't know what they translated as either. She only knew them in the Burmese language. So I just put them in and they've... it's not bok choy, but it's very similar to bok choy. And it's what I used instead of cabbage in my soup yesterday [Laughter]. This soup—it's not Delia Smith's ingredients at all.

The gardeners I encounter regularly plant things shared with them and always manage to use the produce or find a home for it. This encourages a taste for adaptability that is responsive and flexible. As Greta notes, while she used a recipe to prepare one of her large batch soup concoctions, the make-up of the final product bore little resemblance to the original.

This adaptability was also evident in the oft-encountered reluctance of gardeners to 'weed out' or remove food-producing plants that had unexpectedly grown, or which had grown in unintended places. People often spoke about 'letting the plant be' or 'letting it do its thing' even when the plants interrupted their gardening plans. This was not always the case; sometimes plants in the wrong place were a source of great frustration. It was quite common for potatoes to be constantly popping up for years after they were first grown to break-up the clay soils often encountered in Canberra, interrupting other plans for the beds. However, these interruptions were also regularly greeted with excitement, as a bonus for the garden where, as one community gardener noted, 'The soil always comes to the party' (Pamela). These encounters were most often engaged with in a playful, tinkering manner. They involve entering into unexpected relations with others that lead to surprising outcomes requiring ongoing responsive interactions. These are not consistently marked by forms of attachment (gardeners are often quick to point out the plants they did not want growing in particular places) nor outright detachment. They let them stay and care for them even when they wish something else was growing and they could just pull them out. As we saw with Helen's Jerusalem artichokes, the growers always find something to do with the unexpected produce. This motivation is not easily attributable to attachment nor detachment; indeed, both conceptions are unable to adequately encapsulate the relations informing these encounters. But they are marked by forms of togetherness. This togetherness, which is focused on the generative potential of relations not restricted to individual, bounded entities, I suggest, can be conceptualised as togetherness-in-relation that speaks to an enactment of convivial dignity which, through its attention to mutual vulnerabilities, affords an openness to being moved by these relations.

Conclusion

Taste in the gardens we have dwelled in throughout this chapter has been shown to be produced through a myriad of relational encounters among the past, present and future and between human and more-than-humans, bounded and otherwise. Through multisensorial, embodied interactions, the gardeners we have met demonstrate an understanding of taste as generated by relational flows. These productive encounters stretch spatio-temporalities, particularly through notions of nostalgia, and push at the limits of anthropocentric modes of representing our necessary interdependencies. These modes of relation do not fall neatly into dominant categories of attachment or detachment but present as shifting interactive forms of togetherness marked by openness to being moved by the various flows of life that encourages practices of making-do.

As introduced in the preceding chapter, I conceptualise these experiences as togetherness-in-relation and have explored how, in these urban and suburban gardens, these modes of encounter with taste are developed through playful practices marked by tinkering and guided by convivial dignity. The interactions taking place are open to the surprise, and becoming-with, of taste that brings mutual vulnerabilities to the fore. In these responsive ways of being and doing, the behaviours and practices of the gardeners resist anthropocentric representations of human mastery over nature. Yet the gardeners are shown to struggle to represent these. To support and amplify these non-athropocentric modes of being and doing we need to engage in material-semiotic resistance. The enactment of playful modes of tinkering and the conceptualisation of convivial dignity through taste in food producing gardening represents an attempt to push beyond normative anthropocentric limits to support and amplify the ontological foundations for development of the skills and capacities necessary for living well together in times of uncertainty. The multiple affordances of taste to support these new practices of world-making also extend beyond gardens. I follow these broader potentialities in the next chapter through exploration of how these can take shape in the way food flows through the homes of people who shop at AFNs. Here, the affective force of taste is identified as critical to the development of recognition of togetherness-in-relation through attunement to taste and playing with food.

References

Atchison, J. (2015). Experiments in co-existence: The science and practices of biocontrol in invasive species management. *Environment and Planning A*, *47*(8), 1697–1712.

Atchison, J., & Head, L. (2013). Eradicating bodies in invasive plant management. *Environment and Planning D: Society and Space*, *31*(6), 951–968.

Bennett, J. (2010). *Vibrant matter: A political ecology of things*. Durham, NC: Duke University Press.

Blecha, J., & Leitner, H. (2014). Reimagining the food system, the economy, and urban life: New urban chicken-keepers in US cities. *Urban Geography*, *35*(1), 86–108.

Brice, J. (2014). Attending to grape vines: Perceptual practices, planty agencies and multiple temporalities in Australian viticulture. *Social & Cultural Geography*, *15*(8), 942–965.

Buller, H. (2013a). Animal geographies I. *Progress in Human Geography*, *38*(2), 308–318.

Buller, H. (2013b). Individuation, the mass and farm animals. *Theory, Culture & Society*, *30*(7–8), 155–175.

Buller, H. (2016). Animal geographies III. *Progress in Human Geography*, *40*(3), 422–430.

Carolan, M. (2016). Adventurous food futures: Knowing about alternatives is not enough, we need to feel them. *Agriculture and Human Values*, *33*(1), 141–152.

Cloke, P., & Jones, C. (2001). Dwelling, place, and landscape: An orchard in Somerset'. *Environment and Planning A*, *33*(4), 649–666.

De Silvey, C. (2005). Cultural geography: The busyness of being "more-than-representational". *Progress in Human Geography, 29*, 83–94.

Gibson-Graham, J. K. (2008). Diverse economies: Performative practices for other worlds. *Progress in Human Geography, 32*(5), 613–632.

Gibson-Graham, J. K. (2011). A feminist project of belonging for the Anthropocene. *Gender, Place and Culture*, *18*(1), 1–21.

Goodman, D., & Goodman, M. K. (2009). Alternative food networks. In R. Kitchin & N. Thrift (Eds.), *International encyclopedia of human geography* (pp. 208–220). Oxford: Elsevier.

Gröning, G. (1996). *Politics of community gardening in Germany.* Paper presented at the 1996 Annual Conference of The American Community Gardening Association (ACGA), 'Branching Out: Linking Communities Through Gardening', Montréal, Canada. http://www.cityfarmer.org/german99.html

Haraway, D. (2008). *When species meet.* Minneapolis & London: University of Minnesota Press.

Head, L. (2012). Conceptualising the human in cultural landscapes and resilience thinking. In T. Plieninger & C. Bieling (Eds.), *Resilience and the cultural landscape: Understanding and managing change in human-shaped environments* (pp. 65–79). Cambridge, UK: Cambridge University Press.

Head, L. (2014). Contingencies of the Anthropocene: Lessons from the 'Neolithic'. *The Anthropocene Review*, *1*(2), 113–125.

Head, L., Atchison, J., & Phillips, C. (2015). The distinctive capacities of plants: Re-thinking difference via invasive species. *Transactions of the Institute of British Geographers*, *40*(3), 399–413.

Hitchings, R. (2007). How awkward encounters could influence the future form of many gardens. *Transactions of the Institute of British Geographers, New Series, 32*, 363–376.

Hugner, R. S., McDonagh, P., Prothero, A., Scultz, C. J., & Stanton, J. (2007). Who are organic food consumers?: A compilation and review of why people purchase organic food. *Journal of Consumer Behaviour*, *6*(2–3), 94–110.

Lawson, L. J. (2005). *City bountiful: A century of community gardening in America.* London: University of California Press.

Mol, A. (2010). Actor-network theory: Sensitive terms and enduring tensions. *Kölner Zeitschrift für Soziologie und Sozialpsychologie Sonderhefte*, *50*(1), 253–269.

Norberg-Hodge, H. (1991). *Ancient futures: Learning from Ladakh.* San Francisco, CA: Sierra Club Books.

Pincetl, S. (2007). The political ecology of green spaces in the city and linkages to the countryside. *Local Environment*, *12*(2), 87–92.

Puig de la Bellacasa, M. (2017). *Matters of care: Speculative ethics in more than human worlds*. Minneapolis, MN: University of Minnesota Press.

Seyfang, G. (2005). Shopping for sustainability: Can sustainable consumption promote ecological citizenship? *Environmental Politics*, *14*(2), 290–306.

Seyfang, G. (2006). Ecological citizenship and sustainable consumption: Examining local organic food networks. *Journal of Rural Studies*, *22*(4), 383–395.

Seyfang, G., & Haxeltine, A. (2012). Growing grassroots innovations: Exploring the role of community-based initiatives in governing sustainable energy transitions. *Environment and Planning C: Government and Polic*, *30*(3), 381–400.

Steel, C. (2009). *Hungry city: How food shapes our lives*. London: Vintage.

Turner, B., & Henryks, J. (2012). *A study of the demand for community gardens and their benefits for the ACT community*. Canberra, ACT: ACT Government. Retrieved from https://www.planning.act.gov.au/topics/current_projects/research-based-planning/demand_for_community_gardens_and_their_benefits

Turner, B., & Hope, C. (2015). Staging the local: Rethinking scale in farmers' markets. *Australian Geographer*, *46*(2), 147–163.

Whatmore, S. J. (2002). *Hybrid geographies: Natures cultures spaces*. London: Sage.

Whatmore, S. J. (2013). Earthly powers and affective environments: An ontological politics of flood risk. *Theory, Culture & Society*, *30*(7–8), 33–50.

5 Taste in shopping

Introduction

In this chapter, through a focus on alternative forms of place-based, direct food exchange—broadly referred to as Alternative Food Networks (AFNs), though sometimes known as alternative agri-food networks or AAFNs (Andreé, Dibden, Higgins & Cocklin, 2010), I build on the previous chapter's identification of the affordances that encourage people to be moved by food in gardens to explore how affective atmospheres can support attunements to taste. These attunements are then shown to be capable of encouraging the development of responsive relations of togetherness predicated on convivial dignity. In so doing, I reinforce the necessity of identifying and representing alternative forms of togetherness by extending my analysis of the problems with the invocation of narratives, practices and ethics of care and attachment as a means of prompting engagement in more ecologically sustainable practices, particularly in relation to food.

Much of the existing literature on AFNs tends to perpetuate anthropocentric narratives within which humans can make active choices—even if constrained by structural inequalities—to exercise care for themselves, their family, their community (including farmers) and the environment through their decision to engage in forms of 'ethical consumption'. While such forms of care can have positive impacts on the wellbeing of some of those involved in these relations, they can also reinforce normative understandings of a human/more-than-human divide. When care is identified as the raison d'etre for engagement with AFNs, a hyper-separation of humans is in danger of being maintained through pleas to make a benevolent choice to treat the nonhuman well. Furthermore, for those lacking a green ethic care is unlikely to provide a promising basis for motivating environmental sensibilities and low-resource use. Indeed, competing hierarchies of care may see these behaviours sidelined in favour of attention to more immediately sensed concerns (such as caring for their own bodies and those they love).

While elements of anthropocentric narratives can motivate and permeate AFN shopping behaviours, this chapter draws on ethnographic data to argue that through a focus on the materiality of foods and the relations that enable them to take form as manifest in their places and tastes, AFNs can also enable consumers to rethink and redo their subjectivities and relationships with food and the food

DOI: 10.4324/9780429424502-5

system. This can occur through consumers becoming more attuned through embodied, sensorial engagement with food—a training of their sensitivities (Mol, 2010b) to the uncertainty of human and more-than-human elements—from seasonal fluctuations to gut flora—that contribute to taste and tastes. Through these processes, awareness of mutual vulnerabilities are brought to the fore and the foundations may be laid for more-than-humans to enter the ethical arena by prompting growing awareness of the interconnectedness of humans and nonhumans in a broader, intimate ecological web. In the snippets of food stories discussed throughout the chapter, we hear the stirrings of alternative places and practices from which we can initiate new ethical imaginings of togetherness. These modes of togetherness are not induced by care, but speak to alternative relational forms that, I suggest, can be represented as being guided by convivial dignity. These alternatives, and the playful mode in which they can be realised, are explored in this chapter through the generative opportunities of attunement to senses and the force of affective atmospheres in AFNs and their capacities to contribute to forms of mutual adjusting and adaptability among humans and more-than-humans revolving around food. Such 'tunings' (Carolan, 2011, 2015), which support a shift away from seeing mutually entangled relations as being predicated on a need to 'care' for others, may be capable of encouraging the development of modes of being and doing better able to attend to relational forms of living well together.

Caring in alternative food networks

Alternative Food Networks (AFNs) have grown considerably over the last couple of decades in the minority world. These forms of food distribution most notably include the flagship (Brown & Miller, 2008, p. 1296) farmers' markets as well as community supported agriculture (CSAs) schemes. The latter most regularly manifests in schemes whereby participants pay for fruit and vegies in advance and, where food is successfully grown, they then receive a regular delivery of produce. CSAs enable consumers to share the risks associated with food production with the farmer. Paying up front enables investments to be made and provides some security to the producers, regardless of crop success. Farmers markets are also seen as benefiting farmers with direct selling, cutting out the middlemen and thus increasing farmer profits.

However, growing participation in these alternative forms of food provisioning has not only been fuelled by concern for the farmers. Consumers are increasingly attentive to the health, environmental and economic impact of their food purchasing decisions (Baker & Burnham, 2001; Byrne, Toensmeyer, German & Muller, 1991; Goodman & Goodman, 2009; Harper & Makatouni, 2002; La Trobe, 2001; Paxton, 1994; Pretty, Ball, Lang & Morison, 2005; Rosegrant & Cline, 2003; Swinnen, 2007; Wandel, 1994). Motivating many of these food provisioning shifts is the belief that the contemporary, mainstream food system underpinned by international agri-business is a key contributor to negative health, environmental and economic impacts (Baker & Burnham, 2001). As such, a turn to local AFNs to source food is often viewed as a direct challenge to the modern food system (an expression of a wish for greater choice and control over the food ingested) and a desire for 'reconnection' (Brown & Miller, 2008; Morris & Buller, 2003; Sonnino & Marsden, 2006; Turner & Hope, 2015). Along these lines, Seyfang sees 'localised food supply chains' as constituting a form of 'alternative sustainable consumption' which 'aim[s] to strengthen local economies against dependence upon external forces, avoid unnecessary global food transportation (cutting 'food miles') and reconnect local communities with farmers and the landscape' (Seyfang, 2005, p. 300). These are practices often represented as forms of care.

Care is a common motif employed in representations of AFN participation. As indicated above, this includes care for producers through improving the economic well-being of local farmers and, through them, contributing to the economic well-being of the local community at large (Alkon & McCullan, 2011; Renting, Marsden & Banks, 2003; Zepeda & Li, 2006). Environmental care is also commonly linked to motivating AFN participation. This care is seen to be manifest in the focus on consuming local foods with an emphasis on reduced food miles as well as being evident in beliefs and claims that the foods being consumed are produced through practices that are more responsive to the land and broader ecological conditions, usually assumed to involve employment of organic or low-input farming techniques enabled through less intensive production (Coit, 2008; Morris & Buller, 2003; Seyfang, 2005). This is regularly supported by claims of local producers, many of whom offer opportunities to visit farms to allow consumers to 'see' it for themselves. Provisioning food through these networks is, thus, often understood as a way for consumers to care for themselves and immediate family and friends as well as place (inclusive of broader environmental concerns).

Care and privilege in 'good' food politics

Of course, such a rosy picture of AFNs does not encapsulate the messiness of our lived realities and I will once again be at pains throughout this chapter to avoid my work being linked to assumptions about 'good food' and how to go about 'bringing this good food to others' (Guthman, 2008a, p. 434) that are so often aligned with these networks. I also wish to distance the arguments developed here from claims that the so-called capacity for AFNs to 'reconnect' people to food and producers will necessarily induce positive responses and prompt behavioural shifts. I share the wariness indicated by Guthman when she writes:

> ...'knowing where your food comes from' has become one of the most prevalent idioms of the current agro-food movement in the US, as if awareness of the intimacy of food will automatically propel one to make reflexive, ethical food decisions.
>
> (2008b, p. 1175)

The notion that local food enables better care for self, others and the environment is highly questionable. Indeed, the concept of 'local' itself has been shown to be deployed and understood in a multiplicity of ways (see Turner & Hope, 2015). Some research indicates that consumers incorrectly assume that food sold at farmers markets is organic and thus purchase is based on their associated assumptions of health and environmental benefits (Radman, 2005). Other investigations into food miles have demonstrated that the car trips consumers undertake to provision food are collectively greater than the carbon emitted by long-haul transport (see Saunders & Barber, 2008). This has been shown to be true of Canberra consumers with frequent car trips to buy food demonstrated to have a greater impact on carbon footprint than the transport choices made by large corporations trucking food into the city (Ryan, 2011).

The privileging of the contested notion of the 'local' within AFNs has also been heavily critiqued. This is exemplified by Hinrichs' (2003) notion of 'defensive localism' which is generated when goods and people identified as being local are represented as distinctly better (in relation to a whole raft of areas—social, environmental, economic) than non-locals. Thus, for Hinrichs, 'defensive food system localization tends to stress the homogeneity and coherence of "local", in patriotic opposition to heterogeneous and destabilising outside forces, perhaps a global "other"' (Hinrichs, 2003, p. 37). Those working within a defensive localism framework would contend that participation in consumption of these goods enables provision of better care for self, others and place. Even when these forms of care are stretched to encompass global ethical concerns, only particular people and places fall within the purview of care as Guthman (2011, p. 198), reworking the ideas of Allen et al. (2003) states: 'Localism bounds the world to be cared for'. Guthman also notes:

> Even short supply chains that are international in scope, such as Fair Trade, put coffee-producing peasants in Oaxaca inside of moral reach while those in, say, the Gambia, where no Fair Trade commodities are produced, are made even more invisible.
>
> (2008b, p. 1177)

AFNs have been rather convincingly shown in some settings to privilege the lives and livelihoods of some over others. That is, they facilitate the enactment of care for certain segments of the population often by obfuscating the suffering—or at the very least the lack of care—for others. This is evident in Guthman's (2003, 2008a) work on organic food and farmers' markets and in Slocum's (2007) work on sustainable farming and food security in places such as farmers' markets and co-ops throughout California. In these studies, particular forms of privilege are identified as common to participants in these AFNs. Efforts to ensure farmers receive 'decent prices' has contributed to the placing and shaping of these alternatives as the purview of 'relatively well-off consumers' (Guthman, 2008a, p. 431). The framing of these sites as 'white spaces' (with some exceptions, as noted by Alkon & McCullan, 2011) has also been emphasised (Guthman, 2008a; Slocum, 2007). So, while we regularly encounter representations of lower socio-economic classes as those making bad food choices and being in need of education to encourage consumption of 'healthier' foods, access to alternatives remain structurally, physically and perhaps sensorially limited. These notions reinforce the idea that local food systems do not exist in isolation from broader social, economic and ecological pushes and pulls and cannot be understood as simply promoting more beneficial and ethical relationships among people and the food system. The realities are far more complex. But acknowledging these complexities and limitations is not the same as claiming that AFNs are only able to reproduce detrimental inequalities.

While I do not wish to discount a critical approach to AFNs, such analyses should not obfuscate the potential openings for alternative ways of being and doing that may also be enacted through these forms of food exchange. As Slocum writes, 'Farmers' markets are spaces where the stickiness of whiteness as well as hopeful interactions across difference may be apparent' (2007, p. 529). It is this potential for 'hopeful interactions', stretching to include more-than-humans such as the food itself and the climatic and soil conditions that enable its production, that I wish to explore throughout this chapter. As we have seen, some suggest that this potential revolves around notions of care and the caring practices that can be enacted within AFNs. McCarthy, for example, notes that 'the goal of these alternative networks seems to be to create spheres of capitalist production and exchange that are more caring, but not necessarily external to capitalism' (McCarthy 2006, in Guthman, 2008b, p. 1176). However, as identified in earlier chapters, I contend that care is a very limited form of action that is too closely entwined with notions of attachment and connections based on intimacy and familiarity, the formation of which may not always be possible nor desirable. As such, here I flesh out my arguments for alternative ways of conceptualising and narrativising 'doings' and 'beings' with the world through the formation and expression of taste in AFNs. Rather than care, here I explore how practices of play enacted with convivial dignity may lay the foundations for more generative practices of human and more-than-humans living well together. These are practices I have caught glimpses of in everyday food habits of the participants in my research.

Problematising the caring citizen-consumer

The links between care and market relations are particularly complex. While ethical consumption and the concomitant rise of the citizen-consumer is celebrated by some as a means for individuals to bring about change, for others it simply reproduces existing inequalities while salving the conscience of a few. Lewis and Potter err on the side of the former noting that practices of ethical consumption can extend beyond being 'a panacea for middle-class guilt' (Littler, 2011, p. 27):

> Whether through injunctions to buy 'guilt-free' Fair Trade chocolate, to minimize the consumption of energy and water on behalf of the planet, or to recycle or swap goods as a means of reducing consumption overall, mainstream consumer choice is increasingly marked by questions of 'care, solidarity and collective concern'.
>
> (Lewis & Potter, 2011, p. 4)

Here, 'care, solidarity and collective concern' are presented as generative concepts capable of effecting change. The risk remains, however, that collective action will be predicated on forms of care centred solely on humans making active choices, responding to the needs of some (likely to be those with whom they configure loose attachments or forms of identification) over others. The rise of the food citizen (Lockie, 2009, p. 194) brings some of these risks to fruition.

Lockie identifies two key forms of food citizenship, both of which focus on the potential of consumer choice. Firstly, there is one rooted in ecological empathy with a local community focus and, secondly, one ensconced in the self-regulation and surveillance that produces responsibilised citizens (Bauman, 2008, p. 27) typical of neoliberalism. According to Lockie:

> Despite their differences, both ecological and neoliberal models of citizenship proffer a future in which people take responsibility for the consequences of their consumption behavior. No longer the passive recipients of whatever the food industries supply, 'food citizens' must act reflexively and proactively to re-invent for themselves their identities and practices as food consumers.
>
> (2009, p. 200)

These food citizens are positioned as being able to exercise their 'care, solidarity and collective concern' (Barnett, Clarke, Cloke & Malpass, 2005, p. 45) for self, others and the environment through deliberate and careful purchasing behaviours.

On the flipside, there are those who argue that because AFNs remain largely circumscribed within neoliberal practices, they offer little hope for a rethinking and redoing of food systems. For some, the focus on individual action within the concept of the 'food citizen' works against the very possibilities of change. Drawing on the work of Gabriel and Lang, DeLind quotes:

> What this vision of...citizenship lacks...is any wider notion of social solidarity, civic debate, coordinated action or sacrifice. It individualizes the idea of citizenship, as if becoming a citizen is a matter of individual choice alone. In this way, citizenship becomes a lifestyle, however praiseworthy and necessary, which can easily degenerate into tokenism and is hardly likely to alter the politics of consumption.
>
> (Gabriel and Lang, 1995, p. 182 in DeLind, 2002, pp. 218–219)

Of course, within all of these frameworks little consideration is given to the materiality of the food itself. Through these discourses of citizenship and ethical consumption food tends to be positioned as 'objects awaiting cultural inscription' which, even if purchased within the parameters of care, are still 'objects' that 'exist for us; their sole purpose being to be bought, consumed, enjoyed or put to use-by humans' (Turner, 2014). In this way, it may not be the forms of relations of monetary exchange that should be our sole focus when analysing the roles (past, present and future) of AFNs but the relations, and their affects and effects, among the entities themselves and the possibilities for reconfiguration that these offer. Practices of care may not be best suited to shifting these relations.

The material limitations of care

A focus on the instrumental aspects of care, as outlined above, has tended to elide the embodied, visceral encounters with food that can occur through ethical consumption, including in AFNs. The materiality of food and the relations that enliven it matter here. For some, it is this that keeps them away from such sites (as in Guthman's discussion of people who desire certain foods and wish to be able to shop 'with anonymity, convenience and normality' (2008a, p. 443)). Whereas, DeLind (2006) argues that a focus on the economics of AFNs, when viewed as enrolling people in a form of market-relations in which those who see themselves as 'wise consumers' can make 'another delightful, and possibly guilt-assuaging, choice', is unlikely to generate significant changes:

> [b]y narrowing their focus to the rational and the economic, [local] movement activists tend to overlook (or marginalize) the role of the sensual, the emotional, the expressive for maintaining layered sets of embodied relationships to food and to place.
>
> (DeLind, 2006, p. 121)

This is largely achieved by positioning citizens as somewhat passive consumers of food divorced from food's materiality, from production to consumption. Indeed, De Lind goes on to observe in relation to local food that 'Without an emotional, a spiritual, and a physical glue to create loyalty, not to a product, but to layered sets of embodied relationships, local will have no holding power' (DeLind, 2006, p. 126). Perhaps what is needed is not hold-fast glue but relations that enable forms of togetherness that are able to flex, adapt and move in response to encounters among human and more-than-humans. These relations may not be best supported by, or represented as, forms of care.

In the discussion of AFN consumers in Canberra developed throughout this chapter, we will see that, while shoppers do employ expressions of care when talking of their food provisioning practices, there are also other things going on that may offer alternative bases for generating a reconfiguration of human/more-than-human relations. I suggest that these occur through practices of playful experimentation that enable the development of relations of togetherness marked by experiences of convivial dignity. These are not neat, easy relations but ones that can involve discomfort and trade-offs.

Food shopping is rarely carried out in a consistent manner with practices directly matching stated concerns and beliefs. Instead, it is often marked by trade-offs between 'affordability, accessibility and ethics' (Seyfang, 2007, p. 130). But not all consumption is necessarily carried out in a manner that involves conscious assessing of trade-offs. When bringing into view the embodied, visceral encounters with foods themselves of the participants in this research, we can detect an openness to materiality that induces responsive practices that may enable more-than-humans to enter into the ethical arena. This tends to be accompanied by growing awareness of the interconnectedness of humans and more-than-humans in a broader, intimate ecological web. As we have seen in the gardens in the previous chapter, particular contextual affordances (which can stretch spatio-temporalities beyond the present) can help sharpen this responsiveness or capacity to be moved by nonhumans. While this can prompt adoption of an ethic of care, it can also lead to other kinds of engagement marked by togetherness that fray the edges of anthropocentric narratives of human exceptionalism. Play, as a mode of interaction marked by mutual adjusting, is one way of conceiving of these processes. The notion of play I employ is not always whimsical or care-free; it can be hard work, particularly when enacted with convivial dignity, as it is marked by risks and vulnerabilities most evident in its divestment of control, which challenges anthropocentric norms. Shopping for fresh produce through AFNs may well be a key site where responsiveness can be sharpened and play with convivial dignity enacted. To explore these possibilities, I now turn to research with 25 Canberra-based self-identified AFN shoppers.

A taste for AFNs

All the shoppers in this research purchased items from both AFNs and mainstream supermarkets. While the majority avoided buying fresh fruit and veg from supermarkets, this did occur on the odd occasion. Taste, as we saw in Chapter 4, is both socially and materially produced. The bodily sensations and feelings induced by gustatorial taste do not act alone, but these bodily sensations and the materiality inducing them have been given scant attention in the literature. Throughout this chapter, I draw on discussions with people who identify as AFN shoppers to explore how these bodily sensations of taste within particular geographical locales and consumer settings impact both on what people 'do' in their everyday food practices and how they conceptualise these.

The conversations with AFN shoppers were carried out in a semi-structured manner with an interview guide that began by asking people to reflect on where they had shopped for food, what produce they had bought and how they had used it across the previous week. After the interviews, the shoppers were invited to keep a diary for a period of one to four weeks where they recorded their food habits and observations. No one was asked to comment explicitly about tastes or bodies in either of these forms of documentation, but in both they emerged as key ways of conceptualising food experiences ranging from the way bodies moved in shopping sites in response to the material 'qualities' of food to the ways bodies were moved by the ingestion of certain foods. These encounters coalesced around the capacity of sensorial, embodied experiences and the force of affective atmospheres to support attunement to the human/more-than-human entanglements that produce taste and how particular configurations of these can support adaptability and agility in tastes marked by flexible 'tunings' and 'retunings' (Carolan, 2011, 2015). While these themes were identified across the participants, in this chapter, after an initial introductory discussion, I use the food stories of specific individuals to provide a more fine-grained approach to the complexity and affective force of the food experiences of those contributing to this research. To lay the foundations for this approach, we start with the role of 'place' in these encounters.

Sensorial, embodied tastes: stretching affective atmospheres

Much has been written about the social experience of AFNs, particularly farmers' markets, with the very modes of staging of these events often identified as central to their appeal. The discussion in this section explores the relationship to place in a broader sense by looking at 'affective atmospheres' (Anderson, 2009) that are not contained to specific sites on the day of shopping but which stretch out to ongoing shifts and movements which, as Anderson writes, 'exceed the assembling of bodies' and are marked by 'tensions' (2009, p. 78) among 'presence and absence, materiality and ideality, definite and indefinite, singularity and generality' (2009, p. 80). My interest is in the affective force of the visceral, embodied experiences of the AFN shoppers. As Deleuze and Guattari write, 'affects are becomings' (1987, p. 256). Drawing on Haraway (2008), I would suggest affects, and affective atmospheres too, may be best understood as becomings-with. Attunement to these in the case of AFN shopping contributes to the shaping of 'tastes' and the development of a 'taste' for becomings-with. This influx nature of becomings means that affective atmospheres are not stable nor controlled by, or contained within, any specific elements. Rather, they are produced through relations among numerous human and more-than-human entities. As Anderson writes:

> Returning to Deleuze and Guatarri, we can say that atmospheres are generated by bodies—of multiple types—affecting one another as some form of 'envelopment' is produced. Atmospheres do not float free from the bodies that come together and apart to compose situations. Affective qualities emanate from the assembling of the human bodies, discursive bodies, non-human bodies, and all the other bodies that make up everyday situations.
>
> (Anderson, 2009, p. 80)

In this way, atmosphere can be seen to be produced through a form of play among various entities, within which their very boundedness can be called into question. The variable outcomes are developed through responsive practices when in play and thus cannot be predetermined. While atmosphere can work in this way to induce 'collective affects' these are also experienced differently by different bodies in varying contexts. Bodies and materialities (human and more-than-human) and the relations that enliven these and form among them are key to generating these becomings-with through modes of playful experimentation—but what is generated exceeds a simple sum of the parts. Atmospheres are shared as well as deeply personal, as Anderson writes:

> . . .they are impersonal in that they belong to collective situations and yet can be felt as intensely personal. . .atmospheres are spatially discharged affective qualities that are autonomous from the bodies that they emerge from, enable and perish with. As such, to attend to affective atmospheres is to learn to be affected by the ambiguities of affect/emotion, by that which is indeterminate, present and absent, singular and vague.
>
> (Anderson, 2009, p. 80)

These tensions permeate the experiences of those involved in this research and I attempt to map these out by beginning with discussion of the affective force of particular shopping sites before moving into the affective force of taste on particular bodies

The 'feel' and affective force of shopping sites

The non-narrativisable yet sensorially powerful affects of provisioning food through AFNs was identified by all participants in this research. All shoppers spoke of the 'feel' of the markets or retail outlet as being a key motivator for consuming at these particular sites. When speaking about this, they commonly contrasted the 'feel' of the AFN provisioning experience with that of mainstream supermarket shopping and expressed this in terms of 'feelings', mobilising emotion-laden language. AFN shopping was regularly described as being highly pleasurable, whereas supermarkets were 'horrible' places that, while knowable and translatable across settings, lacked any particularly unique characteristics. 'A supermarket's a supermarket', as Julie said while another participant, Judy, commented:

> At the growers' market, you know, you can just waft around, be so calm and take your time and you feel very calm and enjoy all the sights and sounds and smells and colours, and in the supermarket usually I just want to get out as quick as I can, get in and out, and, yeah.

The speed of consumption within supermarkets was spoken about as producing an unsettling and disembodied experience. As Anja notes:

> But, you know, sometimes I go to the supermarket and I completely just turn into a zombie, like quite mindlessly going up the aisles, you know its autopilot now I know what [laughs], you know the things that go in the

> basket after 15 years, week after week after week. So you don't actually have to think at all about your shopping experience, it can be just this kind of mechanical action.

The affective responses to both settings were equally passionate. Supermarkets were 'hated' and seen as 'horrible' while AFN sites were 'loved' and 'uplifting'. While the previous quote indicates the drudgery of routine, the majority of participants also followed regular routines in their AFN shopping, developed via 'trial and error' over time through trying different produce and producers. Once routines were established, this typically involved going at the same time each week, taking the same route through the market or shop, purchasing from the same stalls and often buying many of the same items week in and week out (particularly dairy products, bread, fish and meat). The routines usually only altered in response to seasons when new producers and produce arrived and were likely to sell out quickly or when sellers were moved about the markets to different locations. Despite these quite routine approaches to both conventional and AFN sites, the affective atmosphere of the AFN shopping contributed to (as well as being produced by) a heightened attunement to other bodies and non-humans in these spaces.

Responsiveness at AFNs appeared to be sharpened partly as a result of the affective atmospheres of these sites, which combined in various ways through affective entanglements to induce affects of joy and pleasure that were not easily described nor able to be directly attributed to specific entities. As one participant, Mary, noted of shopping at a farmers market retail outlet:

> Yeah, it just feels good in there. . .there's no one in a hurry, and it's just. . .I know the money's. . .and I know the money's going to good people, you know people who are trying to earn a living. And look, I feel safe.

This good feeling cannot be solely attributed to Mary's desire to shop 'alternatively' and support the growers and shop owners because the shop also made her 'feel safe'. She was unable to identify why this was, but it was a powerful, embodied feeling. Another participant, Alicia—a mother with two teenage children who shopped at both the markets and the retail outlet, also observed that she shopped at these sites because, 'You know, you come out, you feel good. You go, you feel good. You know, you smile, you feel happy'. These 'feelings' were unable to be linked to any specific elements of the experience. Alicia spoke of accessing 'good food' that enabled her to care for her children as well as her body (a motivator here was the fact that she was fond of telling 'everybody I'm going to live to a hundred and unless I treat my body better than it's been treated for the last 40 years, I'm not going to get there'). She also enthused about conversations she had with producers and other shoppers that she believed contributed to the creation of a sense of community that made the markets 'feel good'.

Despite the routines in place in AFN shopping, farmers' markets in particular, were commonly described as being 'vibrant', 'buzzy' places with great 'atmosphere'. They were, I suggest, seen to be playful sites that were not only contrasted with supermarket shopping but also the apparent drudgery of daily working life. As Jenna explains:

> Well, I work in the public service so it's all fairly bland. And then I go to the Growers Market on Saturday and it's a very vibrant place. Everybody, I find, is friendly, the growers that I've been buying from for years I always have a nice chat with them. So it's a very...by the time I come home at... 'cause I go there at 6.30 in the morning, by the time I get home at 8.30 I've had several coffees, said hello to lots of people I know, had chats with friends, got all this good food and it's just a very...I don't know whether I like the word, but kind of uplifting experience. It's a really lovely ritual to start the weekend with. It just starts it off perfectly for me.

This pleasurable experience was sorely missed when people were unable to attend a market. As Kirsty states:

> I have to say that I love it, if I can't go for some reason then I really look forward to the next week because I know the producers and it turns shopping from a drudgery into an energising lovely experience. You're outside, it's either cold or the beginning of the morning, I might see other people around and it turns it into...it makes my life in Canberra more rich, for sure.

Others identified that things didn't 'feel' right for the following week if they weren't able to make it to the markets and had to source food elsewhere.

Stretching affective atmospheres through taste

While the affective atmospheres of the 'event' of AFN shopping are an integral component of every participants' AFN shopping story, this is not confined to the experiences of the duration of the shop itself. Instead, the materiality of the food purchased—its sensorial qualities, most notably 'taste'—maintain connections to the market's affective atmosphere after the food has been taken to homes and stored for the coming week. The force of these affective atmospheres is not simply experienced within the spatio-temporal confines of the event or site of purchase itself but is seen to be embedded in the carefully curated and produced food that is purchased and how this makes bodies feel. As Anja notes:

> you go for the experience and the sort of aesthetics of you know going early to the markets, being around a buzzy place where people are displaying their food to show the best of it and they take some pride in their food...And they value food and you're mixing with people who value the produce that they have produced or that they've transported for your consumption.

The relational impact of these encounters produces affective atmospheres that 'envelop' the shoppers and contribute to developing their attunement to human and more-than-humans and their relational entanglements. Commonly, this was linked to the capacity to engage directly with producers to find out about the foods, the conditions in which they are grown and challenges faced as well as the very material qualities of the foods themselves. There was a great sense of anticipation before attending the markets as shoppers wondered what produce would be available. The vibrancy, excitement and anticipation generated by the markets as well as the specific staging of the event with its 'rustic' modes of presentation and sense of spontaneity were key to the generation of the playful affective atmosphere, as exemplified by Julie:

> ...the rough and readiness of it and the fact that it's not all contrived, that people just come along and set themselves up with their stalls...I just like that. If it became too organised and set up, I'd still go there, but it sort of takes away something.

Taste for the markets was generated by embodied experiences with the affective atmospheres that prompted bodies to respond by engaging in forms of playful tinkering marked by mutual adjusting. This occurs through responsive interactions with humans and more-than-humans and exceeds the confines of the markets themselves. The spatio-temporal nature of affective atmospheres and play are stretched in these sites through the materiality of the food and the way this is handled between markets. Taste is key to this. However, while the force of these affective atmospheres was viewed as highly pleasurable by all AFN participants, these can also belie the tensions and exclusions that enable the development of these affective atmospheres. As we have already seen, access to participation in AFNs has been identified as a problematic issue, with Guthman's (2008a) work highlighting the exclusionary nature of these sites. Before we return to a more intimate focus on issues of bodily taste, I first turn to the story of one specific participant who gives voice to some of the underlying tensions related to access that were hinted at by a number of participants in this research. Her story foreshadows how the spatio-temporal stretching of place through affective atmospheres and its links to the material and sensorial notion of taste could also perpetuate and heighten forms of exclusion.

Accessing a taste for place: Amelia's story

At the time of our discussions, Amelia was a single woman living in a share house in Canberra's inner north. She did most of her fresh produce shopping at the local food co-op, the farmers' retail outlet and the farmers' market (though gradually the former was replacing this). While many participants indicated how infrastructure (such as the farmers' markets beginning or the farmers' retail outlets opening) and their ability to access it (how close they lived, whether they had a car, etc.) significantly impacted on their food provisioning habits, they also hinted at the ways that shopping at these sites seemed to require a certain mode of participation and cultural capital to navigate. I use Amelia's stories here to tease out these tensions because she explicitly identifies these as a potential barrier to participation.

Amelia's insights, as was typical among the participants, compared pleasurable shopping experiences in the AFNs and their affective atmospheres to the unpleasant process of enduring supermarkets. As Amelia stated, 'When I walk into a supermarket, I just kind of feel like a bunny in the headlights and a bit overwhelmed'. This she contrasted with the food co-op:

> when I walk into food co-op, I don't feel like that although I recognise that other people quite easily could if it's not somewhere that you're familiar with. I really like being able to kind of dish up exactly how much of any food that I want and kind of look at that and make a decision about do I want the wholemeal plain flour or the wholemeal self-raising flour or the spelt flour or the kama flour or whatever it is. Those are the decisions I want to make about my food, I don't care if it's the Heinz baked beans or the SPC or which has got the most salt in it, just give me one choice which is a choice that someone's kind of made an intentional decision about that. Yeah so, I mean I really enjoy going to the food co-op and it kind of feels like a bit of a hub so it's nice to see what's going on there, in there.

In these comments, we can see quite clearly the tension between different forms of choice—some are validated as they align with a particular ethical stance while others (including some related to normative nutritional advice) are dismissed as irrelevant for this consumer. This latter form of choice fits neatly within the neoliberal notion of consumption, aided and abetted by the regulatory forms of governmentality via nutritional guidelines and labelling (see Guthman, 2007). While such forms of choice might be those that consumers are encouraged to 'care' about, these do not align with Amelia's taste. Amelia's bodily reactions to foods are bound up in the ethico-political decisions she makes about consumption and she has come to understand that certain shopping places allow her to exercise these better than others; knowledge developed through experimenting with different modes of consumption. Her tastes for these choices make her feel comfortable in her chosen AFNs.

As was common when participants spoke about their shopping experiences in general, Amelia's discussion of the co-op and the farmers' retail outlet tended to revolve around notions of bodily movements and speed. In the AFNs, she found herself slowing down. While she identified the affective atmospheres generated here as enabling her to exercise a form of careful and conscious shopping, this was not a universal positive for her. She spoke of sometimes being frustrated by the slowness of the process and service in these spaces when she just wanted to get in and get out quickly. However, despite these drawbacks she commented that:

> for the most part, it [the co-op] definitely feels like a more real experience. Like when I walk out of a supermarket I always feel a bit like, 'Oh, I've just bought all of this stuff that I didn't really need'. I know the food co-op doesn't feel like that.

While, from an ethico-political standpoint, Amelia expresses the belief that she should embrace the slowness that she felt made food provisioning more mindful and 'real', she was consistently torn between this ideal and managing the other everyday tasks required of her. This tension was not always something she was immediately conscious of, but it was experienced through her bodily engagement. This was particularly apparent when she discussed shopping at the farmers' market in contrast to using the farmers' retail outlet:

> I used to go to the farmers' market a lot more and then switched to Choku Bai Jo [farmers' retail outlet] and I was like 'Oh, it takes like ten minutes and I'm done' whereas the markets it's kind of like this couple of hours experience and Choku Bai Jo isn't and it was interesting to reflect on my reaction to it. I was like, 'It's like a supermarket, it's so efficient that's why', and I was 'Yeah', but it's smaller than a supermarket so I can kind of. . .what I like is being able to walk in and, I guess because I don't shop with a plan or to a menu, to kind of be able to get kind of an instant visual of like what's here. And then to probably buy the things that I normally buy so it feels like a very efficient process. And easy to understand I guess in terms of how it runs it's quite similar to like a Fyshwick market where I [used to] shop a lot so there's not a difficult system there to have to learn.

Once again, Amelia emphasises the importance of knowing how to navigate the shopping spaces and practices within them—somewhat akin to knowing the rules of game before one can play—pointing out that the farmers' retail outlets follow a conventional format that enables her to shop efficiently. This notion of efficiency, particularly linked to speed, seemed to Amelia to be at odds with the way she thought AFN shopping should unfold and she likens the retail outlet not only to other commercial markets but also to supermarkets. This speaks to the assumptions often made about AFN shopping that highlights the 'experience' and the relations that unfold within this form of economic exchange as being key to those who support AFNs. These affective atmospheres, as identified in the previous section, are usually represented positively, yet, as Amelia's experiences indicate, there can also be significant tension between ethico-political ideals and the embodied spatio-temporal realities.

Ethical misalignment within spatio-temporal experiences: guilt, trade-offs and responsiveness

Amelia was much more concerned with the materiality of the produce, its localness and freshness and ease of accessibility rather than the general experience in which the consumption practices unfolded. While she felt that supermarkets were difficult to navigate and caused bodily discomfort, prompting her to rush through the shop, she also saw them as efficient (although too big), a quality she viewed as largely positive. These were qualities she also found in the farmers' retail outlet, places that align more neatly with her ethico-political perspectives. The co-op was well known to her and was a place where she accepted some things took longer, even though these could prompt frustrations. The farmers' market, however, stretched the AFN experience beyond a realm that was manageable despite the ethico-political fit. Amelia indicated feelings of guilt in relation to this, suggesting that her desire to avoid devoting two hours to shopping went against the grain of the committed AFN shopper. Her taste for locally produced, ethically curated foods was out of sync with what she saw should be the correct amount of bodily effort and time one should exert to enact their ethico-political beliefs in practice. However, she offset these feelings of 'guilt' expressing her beliefs through other components of her practices, most notably her responsiveness to foods.

The choices Amelia made about food to fit her ethico-political approaches to produce involved adapting to the food options available. She was highly flexible and responsive in regard to what she would purchase and eat each week. She did not write shopping lists nor did she develop a menu plan indicating responsiveness to seasons and fluctuations in foods in relation to climatic and weather conditions. In response to a question about whether there was anything about the AFN sites she visited that she would like to see change she noted:

> So, I feel like there's nothing really that they would particularly need to change. You know, I adapt my food choices to primarily be able to go to those particular places.

Her modes of provisioning had a significant impact on the tastes available to her and also attuned her to the broader contexts that made certain tastes available. However, this was qualified by her decision to continue to purchase some items that she loved (notably a certain type of cheese) from the supermarket. The desire for the sensorial pleasures of some particular foods disrupted her ethico-political principles but, by and large, her capacity to easily access AFNs made the enactment of these principles possible:

> Yeah, and as the farmers' markets developed, you know, I've kind of been going for as long as they've been around, so then when Choku Bai Jo opened and one of the best things about it is that they're open after work. Like that's just, that to me is total gold. That's awesome; you know something that's open in the evenings just makes so much sense. So, it's also somewhere that I want to support, like I value that they are here providing that service.

The affordances of the AFN scene and Amelia's capacity, or cultural capital, to negotiate these spaces made possible particular forms of provisioning. In so doing, her responsive approach to available foods, and places to procure this, encourage forms of playful tinkering with the mode of provisioning and the foods she consumes. However, this play is tempered by the tensions Amelia encounters between her perception of the need for AFN shopping to be slow and mindful which appeals to her and her desires for quick and efficient consumption that fit around her existing lifestyle. Here, again we may be able to discern a difference between notions of care—commonly associated with something enacted over a period of time—and playful tinkering that can occur in multiple time modalities. Amelia's responsive food practices and tastes, moulded by the availability of certain foods in certain places, can be read as forms of play. But these are ones in which her taste for certain places makes her feel like she isn't quite playing by the rules.

A feel for uncertainty

In spite of the tensions identified in Amelia's stories, embodied, visceral encounters with the affective atmospheres of AFN shopping can (but do not necessarily) prompt the development of attunement and responsiveness to food and the nonhuman and human inputs that produce it. The capacity for this to occur is particularly evident in what Carolan (2011, 2015) refers to as a 'tuning' or 'retuning' of the tastes of those choosing to provision their food from these locations which is expressed through adaptations and flexibility in relation to encounters with food. This is evident in the trial and error development of routines to navigate the spatial and social elements of AFN shopping, but also in the responsiveness to the materiality of the foods themselves. Indeed, the gustatory pleasures induced by food were regularly enthused about in the research interviews and this drive for pleasurable food experiences was a motivator for AFN participation. AFN-provisioned food was consistently identified as being tastier than that purchased elsewhere. The embodied, visceral sensations of the food sharpened the responsiveness of the shoppers, inducing particular bodily actions. As Jenna notes:

> ...the taste of the food that I eat is just unbelievable. You know, I've given people at work... I make this salad and people go 'wow that looks good.' It's just an ordinary salad, there's nothing spectacular about it, you know, so I say, 'Well, try some', and it's just totally different to anything they've ever tried and it's... it might just be tomato, cucumber and some greens, you know.

Taste was often the driver of the produce selected, overriding other concerns that people had such as a desire to avoid pesticides. This was exemplified by Anja when she says:

> ...sometimes I might get organic lemons, sometimes I get organic carrots, it's whatever I bump into first, really, in my way around. Sometimes the

> organics are a bit... not great looking and I don't get them for that reason and they don't taste good sometimes, the inorganic lemons can be better and so I get those because I know they taste better.

While perceptions of better taste were regularly identified as the basis for purchasing decisions, as the shoppers said this out loud and then reflected on this during the interviews, it became apparent that many felt that this may not be considered to be a legitimate justification for their choices. These moments of pause manifested in a tendency to qualify the embodied experience of taste with a more 'acceptable' and 'considered' ethico-political position as highlighted by Mary's comments on why she shopped at AFNs:

> Okay. I think there's probably three big things. One is I like to get local produce, and it's not because I have any other belief except it tastes better.
>
> I mean, all the other stuff (a) I do it for all the other reasons, but basically it tastes better. I do like to support local networks, and that's... so that's another reason. The price of stuff is important, so I don't... you know, it's... as I said I'm on a limited income, so price is really important, and proximity. And also... and the other thing I think is the way a place feels.

Mirroring the limited attention paid to taste and the viscerality of 'gut reactions' in much of the AFN literature, in the conversations carried out for this research, taste alone was rarely considered to hold enough weight to justify the extraordinary lengths some people went to source food through AFNs. This often prompted reflections on the modes of care that participation in farmers markets enables.

While many of the practices shoppers engage in were identified as being driven by certain forms of care (as evident in Amelia's ethico-political tastes), I contend that they may be more productively understood as being underpinned by a less directed form of togetherness, one neither marked by strong attachment nor detachment but by a responsiveness to others and understanding of the generative relational entanglements of our ongoing and shifting relational encounters. Indeed, care in AFNs remains problematic given the privileging of particular forms of relations and reification of the local. Given a number of participants (as exemplified by Amelia) were conscious of this, it seems useful to explore how these conceptions of togetherness among humans and more-than-humans apparent in the narratives and practices of all of the shoppers that contributed to this research (even for the few participants who did not articulate an interest in farmer wellbeing and environmental concerns) can be otherwise conceptualised. Doing so draws our attention back to bodies and taste. As bodies become 'tuned' to the 'taste' of the AFNs, they also seemed to become attuned to relational entanglements that expose mutual vulnerabilities. As we see above, these are bodies that don't 'feel' right unless they are fed with certain foods or with foods provisioned in particular ways.

To further flesh out some of the ways in which visceral engagements with food motivate actions and the possibilities for these to move beyond care to encourage interactions marked by playful tinkering capable of generating non-anthropocentric forms of human/more-than-human togetherness, I draw on Deanne's story. It is worth dwelling with her experiences because at the time of the research, Deanne was rather new to AFN shopping. Hers is a story of food engagements in transition, where practices might congeal or shift in various ways over time. It is a story where the body takes centre stage, responsiveness to which prompts a 'taste' for certain material qualities of food that make her body 'feel good'. In her experiences, a desire for certain tastes not only encourages her to shop in certain ways but is generated by the gut reactions of a 'metabolic body' (Mol & Law, 2004, p. 54) beyond her control. Deanne's food encounters speak to efforts to 'train [her] inner sensitivity' (Mol & Law, 2004, p. 48) to become attuned to her gut through forms of playful tinkering. It is also a story which points to some of the limits of care as being the sole basis for developing more productive ethical relations among humans and more-than-humans.

Storying 'gut reactions' with Deanne

Deanne is a low-income single mother without a car who, at the time of the research, had recently decided to overhaul her and her teenage son's eating habits. Her discussions of food provisioning and ingestion centred on visceral engagement and bodily/gut reactions. Deanne focused on the materiality of food, most notably in regards to how these made her and her son's bodies feel. While this could be perceived to be a form of self-care (though this was not a term she used), and Deanne does seem to suggest that foods 'are media for care' (Harbers, Mol & Stollmeyer, 2002, p. 217), this care is only identified as being able to be enacted through attunement to the needs of the gut. Neither the food nor Deanne are capable of caring alone. It is the relational entanglements and training of skills and capacities to be sensitive to these that makes her body feel certain things and act in certain ways. While this resonates with Mol's (2010a) notion of care, Deanne's conception of care was quite limited. She did not dwell on the potential for her AFN-provisioned food to care for others nor did she focus on the impacts of her modes of consumption on producers or the environment.

While care did not adequately represent Deanne's everyday food practices, these were highly responsive and focused on the relational aspects of food and bodies with primary attention paid to attuning to her gut reactions. This attunement prompted the introduction of a juicing routine and inspired efforts to seek out organic produce. While she described herself primarily as an AFN shopper because each week she shopped at a farmers' retail outlet, and prior to that had frequented a farmers' market, her key concern with the perceived material qualities of food (specifically organic) meant she also sourced produce from mainstream supermarkets. While she did not like the feel of supermarkets, she did not express concerns about the social, economic or political impacts of these forms of food distribution.

For Deanne, a typical week's shopping unfolded as follows:

DEANNE: I normally buy a lot of the greens, like the spinach and the endive, and. . .it depends on the carrots. I try to only buy the organic from there [farmers' retail outlet] if I can; and then if I can't get what I want—so the beetroots and stuff like that, so mainly for my juicing, and my pumpkins and that, and if I don't like the look of that, or they haven't got much there when I go, then I'll just get the organic stuff in Coles; but if I can't get the organic stuff then I'll go to the normal stuff. First priority is always organic.

INTERVIEWER: Can you expand and explain to me what it is about organic that's attracted you towards that?

DEANNE: I just feel better on it. I just feel better. I just feel better when I eat it, I feel better after I've eaten it, I just don't like what they put on the other products.

The organic `attributes of the food were viewed as having an immediate and ongoing impact on the way Deanne feels. Her sensorial response is also linked to assumptions about what is not put on her products. This visceral engagement with the qualities of the food and the feeling of her body motivates particular shopping practices marked by a form of attentiveness to the materiality of the food (both visible and unseen) and her perception of bodily responses to this.

Gut reactions as the locus of her food-provisioning habits also applied to Deanne's broader eating routines which she attempted to keep quite consistent, observing that 'My body just doesn't. . .It knows what it likes and what it doesn't'. Throughout the conversations, Deanne's body was consistently invoked as a materiality beyond her control that required her to become attuned to its likes, dislikes and the need to attend to it in particular ways. When talking about what motivated her juicing routine she stated 'Just my body told me that I needed to do that'. However, these bodily needs were not always easily accommodated. Deanne was very conscious (and rather concerned) about the higher price of organic foods, noting they could cost up to three times as much as conventionally produced foods. This economic burden was keenly felt. While the high cost did not deter Deanne from purchasing organic foods, it did prompt her to begin planting her own garden.

As Deanne talked about her recent eating and shopping experiences, it became possible to read these practices as being marked by a playful, tinkering approach. Her organic preference and juicing routine developed slowly over time, 'bit-by-bit' in response to her gradual attunement to her gut. She shifted from farmers' market shopping to the farmers' retail outlet as this became more convenient for her. The cost of her new habits then prompted interest in growing her own food, with her son about to begin a major composting project to support this. The types of food she was purchasing were also open to tinkering, or adjustment, in relation to seasons and availability, though her organic preference imposed a certain rigidity on her practices. Still, she shopped responsively to 'what looked good' describing her farmers' retail outlet shop routine in the following way:

> So, I generally walk in there, walk to the organic section (and they're always nice and friendly there), and I walk up and down and see what I like the look of, and I stick what I like in my basket.

When asked directly about seasonal availability and its influence on provisioning habits, Deanne had yet to notice any significant impact. However, she saw the farmers' retail outlet, not the supermarkets, as a space that could attune her to seasonal availability prompting her to question the foods (and their mode of storage etc.) that she wished to consume:

> . . .if I can't get it here [farmers' retail outlet], that means it's not in season, so do I want to get it when it's out of season? And how do they keep it when it's out of season? and all those questions that sort of pop up.

These were sentiments directly linked to how her body might respond to these foods. The capacity for attunement to food availability and its particular characteristics (such as appearance and taste) afforded by shopping at the farmers' retail outlet also extended beyond seasons to awareness of the variability encountered in food production including the impact of localised weather conditions and vulnerability to pests and diseases. As Deanne notes:

> . . .I think they can only supply what they can supply; what's growing well, what's not growing well. So, I think it just depends on how everything's going there'cause when we went to the farm [she and her son had been to a farm open day], he [the farmer] said they'd lost a whole lot of one crop, so it obviously depends on what can be supplied. So, I don't think they could really increase their range because I think it's very dependent on what's growing and what's not growing.

While I have suggested elsewhere that exposure to the vicissitudes of farming life facilitated through the direct exchange relation of farmers markets and, by extension, farmers' retail outlets can facilitate forms of connection to people and place beyond the immediate local sphere of concerns of consumers (Turner & Hope, 2014), this is not necessarily always the case. Nor does this embodied knowledge necessarily encourage a form of care for others. In this conversation with Deanne, she is aware of the challenges involved in production and is gradually becoming attuned to seasonal shifts in food supply, but she does not express an ethic of care in relation to these matters. She cares about organics and the way these foods make her (and her son's) body feel. She cares about knowing how her food is produced but she does not outrightly extend this care to others. However, while broader forms of non self-care connection or attachment seems to be missing, Deanne is evidently moved by these others. This is evident in her responsiveness to foods and bodies. She 'listens' to her 'gut reactions' and attributes these responses to the qualities of the food (including their mode of production) and these mould the particular tastes that motivate what and how she eats and where this food comes from.

Taste here is not simply about the sensorial pleasure of foods, but the capacity to induce bodies to respond in particular ways. Through this 'noticing of her body', the relational generation of 'feeling good' was highlighted—body/food/production all work together as matter in relation to induce gut reactions and encourage particular approaches to food provisioning. Awareness of the limits of human control over bodily reactions and food production evident in seasonal availability and climatic/pest/disease impacts on crops was viscerally experienced through Deanne's participation in AFNs. Yet this did not necessarily significantly alter her food provisioning habits nor incite an ethic of care beyond that bounded by her immediate family. It did, however, seem to sharpen her responsiveness to these, possibly laying the nascent foundations for the development of the skills and capacities needed to adapt to uncertain food futures.

While care may not be overtly evident here beyond the immediate family, through her responsive encounters with more-than-humans there is evidence of play—a charge that has the potential to unleash new forms of togetherness revolving around the human and more-than-human components involved in the matter-in-relation of production/food/body. This hint of playful tinkering might represent the beginnings of the retuning of bodies Carolan (2015) suggests is necessary to counteract the training of taste facilitated by the modern industrial food system. Playful tasting attuned to 'gut reactions' may provide one of the places from which we can start to develop and generate the skills and capacities to enact and support these retunings. When this is guided by recognition of mutual vulnerabilities and of the necessity of our togetherness-in-relation, as expressed through convivial dignity, there is potential for this to support more ecologically sensitive modes of being and doing.

A taste for adaptability

Attuning and adapting to elements beyond human control dominate the food stories of the AFN shoppers that participated in this research. This was often associated with an affect of joy as they trained their sensitivities and developed their skills, to respond to the uncertainty of what would be encountered, as well as what would be needed by their bodies and what they would need to do to enable this. These encounters with mutual vulnerabilities tended to prompt playful responses and was seen as key to the AFN experience as one shopper, Mallory noted:

> If you want to have mangoes in July, then you're not going to like it at the Growers Market, but I guess I believe that it's good to eat by the seasons. So, you know, and so you just enjoy it, and yeah, and I think the right food grows at the right time for our bodies, and we're getting out. . . we live in a world that gets us out of touch with that.

Another, Kate, commented of the unknowable elements of the site:

> I don't mind a little variable. I don't mind that sometimes people don't come because they've got some other thing on or whatever, that's the way life is.

This capacity to respond and adapt to human and nonhuman variability was seen to represent the heart of AFN shopping for many. The visceral, sensorial experiences with the AFNs and the foods on offer were the key motivators for the shopping, cooking and eating habits of the participants. They tended to buy seasonal food that 'looked good' and cooked what 'tastes good' or was perceived to be what their bodies needed. These approaches are enlivened by a playful, tinkering approach, insofar as these are embodied, responsive, non-goal-oriented interactions where the modalities of engagement may be familiar but the nature of the players and the potentialities of the encounters are unknown. Sometimes this play didn't work, as exemplified by Mary:

> Actually, last week I did make a disgusting lasagne, that... oh, it was planned for three nights, and I had it once and I thought no, so that went to the chooks.

Being prepared for things to not work and to be able to respond are critical to the risks that we must enter into if we are to engage in the work of being and doing differently with the world.

Conclusion

The potential for playful, tinkering encounters with food, bodies, places and the relations that enliven these in AFNs to support an attunement to taste informed by recognition of the necessity of responsive relations of togetherness—or togetherness-in-relation—has been explored throughout this chapter. This is shown to be enabled through the affordances of the affective force of the matter-in-relation of taste and the affective atmospheres of AFNs that I contend stretch beyond the temporal and geographical limitations of the shopping events themselves. Here, I have also deepened my critique of the use of care as a means of encouraging environmentally sustainable actions suggesting the term has limited potential to disrupt the humanist subject embedded in the renderings of hyper-separation perpetuated in anthropocentric narratives and practices.

I have attempted to demonstrate that in the 'beings' and 'doings' of the AFN shoppers that participated in this fieldwork, we can discern an openness to alternative forms of togetherness supported by practices of playful tinkering which invite participation in risky relations where mutual vulnerabilities are exposed through attunement to taste. This play, I suggest, is often guided by convivial dignity where the necessity of us entering into, and being configured by, shifting relations is recognised. Playful tinkering enacted with convivial dignity offers ways of conceiving of and engaging in forms of living together that don't rely on appeals to relations of attachments through care and which avoid attending to alterity through desires for detachment. Openness and responsiveness to taste is shown to be a collective, relational endeavour that can contribute to the development of the skills and capacities to adapt and adjust to our uncertain food futures.

In highlighting the potential for tastes cultivated through AFN shopping to support adaptation to uncertainty, I have attempted to avoid romanticising these forms of food provisioning. AFNs have been shown to not necessarily be more ethical than other forms of shopping and can perpetuate discrimination and lack of access as well as support labour or farm practices that could be detrimental to the humans and nonhumans involved. However, I have suggested that the capacity to be responsive to the relational flows of matter and life encountered here through taste could provide a basis for rethinking anthropocentric practices and beliefs; attunement to taste can provide opportunities for sensing or 'feeling' sustainability (Carolan, 2015). As Carolan (2015, p. 317) states 'We are also going to need to embrace the fact that it is not enough to know sustainability. We have to literally be able to *feel* it'.

The following chapter builds on this concern with 'feeling' sustainability and the necessary 'training of sensitivities' that can assist displace anthropocentric narratives of human exceptionalism through exploration of how this manifests in a site historically associated with industrial agriculture and embedded within discourses of modernity and human mastery, namely, the Agricultural Show. In this final chapter of the Taste section, I highlight the limits of anthropocentric grammars to conceive of and represent these reconfigurations of the subject and our relational entanglements, emphasising the necessity of material-semiotic initiatives in struggles to promote more ecologically sensitive modes of being and doing with the world. Playful tinkering enacted with convivial dignity is offered as a contribution to support and amplify these reworkings.

References

Allen, P., Fitzsimmons, M., Goodman, M. & Warner, K. (2003). Shifting plates in the agrifood landscape: The tectonics of alternative agrifood initiatives in California. *Journal of Rural Studies*, *19*, 61–75.

Alkon, A. H., & McCullan, C. G. (2011). Whiteness and farmers markets: Performances, perpetuations . . . contestations? *Antipode: A Radical Journal of Geography*, *43*(4), 937–959.

Anderson, B. (2009). Affective atmospheres. *Emotion, Space and Society*, *2*(2), 77–81.

Andreé, P., Dibden, J., Higgins, V., & Cocklin, C. (2010). Competitive productivism and Australia's emerging 'alternative' agri-food networks: Producing for farmers' markets in Victoria and beyond. *Australian Geographer*, *41*(3), 307–322.

Baker, G. A., & Burnham, T. A. (2001). Consumer response to genetically modified foods: Market segment analysis and implications for producers and policy makers. *Journal of Agricultural and Resource Economics*, *26*(2), 387–403.

Barnett, C., Clarke, N., Cloke, P., & Malpass, A. (2005). The political ethics of consumerism. *Consumer Policy Review*, *15*(2), 45–51.

Bauman, Z. (2008). *The art of life*. Cambridge, MA: Polity Press.

Brown, C., & Miller, S. (2008). The impacts of local markets: A review of research on farmers markets and community supported agriculture (CSA). *American Journal of Agriculture Economics*, *90*(5), 1296–1302.

Byrne, P. J., Toensmeyer, U. C., German, C. L., & Muller, H. R. (1991). Analysis of consumer attitudes toward organic produce and purchase likelihood. *Journal of Food Distribution Research*, *22*(2), 49–62.

Carolan, M. (2011). *Embodied food politics*. Burlington, VT: Ashgate Publishing Ltd.

Carolan, M. (2015). Affective sustainable landscapes and care ecologies: Getting a real feel for alternative food communities. *Sustainability Science*, *10*(2), 317–329.

Coit, M. (2008). Jumping on the next bandwagon: An overview of the policy and legal aspects of the local food movement. *Journal of Food, Law and Policy*, *4*(1), 45–70.

Deleuze, G., & Guattari, F. (1987). *A thousand plateaus: Capitalism and schizophrenia* (B. Massumi, Trans.). Minneapolis, MN: University of Minnesota Press.

DeLind, L. B. (2002). Place, work, and civic agriculture: Common fields for cultivation. *Agriculture and Human Values*, *19*(3), 217–224.

DeLind, L. B. (2006). Of bodies, place, and culture: Re-Situating local food. *Journal of Agricultural and Environmental Ethics*, *19*(2), 121–146.

Goodman, D., & Goodman, M. K. (2009). Alternative food networks. In R. Kitchin & N. Thrift (Eds.), *International encyclopedia of human geography* (pp. 208–220). Oxford: Elsevier.

Guthman, J. (2003). Fast food/organic food: Reflexive tastes and the making of 'yuppie chow'. *Social & Cultural Geography*, *4*(1), 45–58.

Guthman, J. (2007). The Polanyian way? Voluntary food labels as neoliberal governance. *Antipode*, *39*(3), 456–478.

Guthman, J. (2008a). Bringing good food to others: Investigating the subjects of alternative food practice. *Cultural geographies*, *15*(4), 431–447.

Guthman, J. (2008b). Neoliberalism and the making of food politics in California. *Geoforum*, *39*(3), 1171–1183.

Guthman, J. (2011). Weighing in: Obesity, food justice, and the limits of capitalism (Vol. 32). Berkely, Los Angeles & London: University of California Press.

Haraway, D. (2008). *When species meet*. Minneapolis & London: University of Minnesota Press.

Harbers, H., Mol, A., & Stollmeyer, A. (2002). Food matters: Arguments for an ethnography of daily care. *Theory, Culture & Society*, *19*(5–6), 207–226.

Harper, G. C., & Makatouni, A. (2002). Consumer perception of organic food production and farm animal welfare. *British Food Journal*, *104*(3/4/5), 287–299.

Hinrichs, C. C. (2003). The practice and politics of food system localization. *Journal of Rural Studies*, *19*(1), 33–45.

La Trobe, H. (2001). Farmers' markets: Consuming local rural produce. *Consumer Studies*, *25*(3), 181–192.

Lewis, T., & Potter, E. (2011). Introduction. In T. Lewis & E. Potter (Eds.), *Ethical consumption: A critical introduction* (pp. 3–23). New York: Routledge.

Littler, J. (2011). What's wrong with ethical consumption? In T. Lewis & E. Potter (Eds.), *Ethical consumption: A critical introduction* (pp. 27–39). London and New York: Routledge.

Lockie, S. (2009). Responsibility and agency within alternative food networks: Assembling the citizen consumer. *Agriculture and Human Values*, *26*(3), 193–201.

Mol, A. (2010a). Care and its values: Good food in the nursing home. In A. Mol, I. Moser, & J. Pols (Eds.), *Care in practice: On tinkering in clinics, homes and farms* (pp. 215–234). Bielefeld, Germany: Transcript Verlag.

Mol, A. (2010b). Moderation or satisfaction? Food ethics and food facts. In S. Vandamme, S. Van de Vathorst, & I. De Beaufort (Eds.), *Whose weight is it anyway? Essays on ethics and eating*. Leuven: Acco Academic.

Mol, A., & Law, J. (2004). Embodied action, enacted bodies: The example of hypoglycaemia. *Body & Society, 10*(2–3), 43–62.

Morris, C., & Buller, H. (2003). The local food sector: A preliminary assessment of its form and impact in Gloucestershire. *British Food Journal, 105*(8), 559–566.

Paxton, A. (1994). *Food miles report: Dangers of long distance food transport*. London: SAFE Alliance.

Pretty, J. N., Ball, A., Lang, T., & Morison, J. I. (2005). Farm costs and food miles: An assessment of the full cost of the UK weekly food basket. *Food Policy, 30*(1), 1–19.

Radman, M. (2005). Consumer consumption and perception of organic products in Croatia. *British Food Journal, 107*(4), 263–273.

Renting, H., Marsden, T., & Banks, J. (2003). Understanding alternative food networks: Exploring the role of short food supply chains in rural development. *Environment and Planning A, 35*(3), 393–411.

Rosegrant, M. W., & Cline, S. A. (2003). Global food security: Challenges and policies. *Science, 302*(5652), 1917–1919.

Ryan, S. (2011). *Buying choices for a more sustainable Canberra: Report for the ACT commissioner for sustainability and the environment*. Canberra, ACT: ACT Government. Retrieved from https://www.parliament.act.gov.au/__data/assets/pdf_file/0003/372171/02b_Buying_choice.pdf.

Saunders, C., & Barber, A. (2008). Carbon footprints, life cycle analysis, food miles: Global trade trends and market issues. *Political Science, 60*(1), 73–88.

Seyfang, G. (2005). Shopping for sustainability: Can sustainable consumption promote ecological citizenship? *Environmental Politics, 14*(2), 290–306.

Seyfang, G. (2007). Growing sustainable consumption communities: The case of local organic food networks. *International Journal of Sociology and Social Policy, 27*(3/4), 120–134.

Slocum, R. (2007). Whiteness, space, and alternative food practice. *Geoforum, 38*(3), 520–533.

Sonnino, R., & Marsden, T. (2006). Beyond the divide: Rethinking relationships between alternative and conventional food networks in Europe. *Journal of Economic Geography, 6*(2), 181–199.

Swinnen, J. F. M. (Ed.). (2007). *Global supply chains, standards for the poor: How the globalisation of food systems and standards affects rural development and poverty*. Wallingford, UK: CAB International.

Turner, B. (2014). Food waste, intimacy and compost: The stirrings of a new ecology? *Scan Journal of Media Arts Culture, 11*, 1.

Turner, B., & Hope, C. (2014). Ecological connections: Reimagining the role of farmers' markets. *Rural Society, 23*(2), 175–187.

Turner, B., & Hope, C. (2015). Staging the local: Rethinking scale in farmers' markets. *Australian Geographer, 46*(2), 147–163.

Wandel, M. (1994). Understanding consumer concern about food-related health risks. *British Food Journal, 96*(7), 35–40.

Zepeda, L., & Li, J. (2006). Who buys local food? *Journal of Food Distribution Research, 37*(3), 1–11.

6 Taste in competition

Introduction

The final chapter in this section pries open the concept of taste a little more to incorporate not only a concern with the relational enactment of its gustatory forms and their effects and affects, but also its aesthetic manifestations. This multifaceted approach demonstrates some of the broader ways in which people can be moved by the relational entanglements of taste while simultaneously drawing attention to the limited capacity of existing grammars—ensconced in anthropocentric conceptions of human-exceptionalism—to capture and represent these modes of interaction. As such, it highlights the need for alternative grammars capable of supporting and amplifying non-anthropocentric ways of doing and being and new ontologies capable of generating more ecologically sensitive practices. In response, I explore the ways that playful tinkering guided by convivial dignity could provide one way of attending to this gap. These arguments are developed through paying close attention to taste in an unlikely site, namely that of the Royal Canberra Agricultural Show.

There are approximately 600 agricultural shows held across Australia each year in small regional towns and capital cities with Sydney's Royal Easter Show being the nation's largest annual event (SRES, 2014) attended by around 900,000 people each year. Typically depicted as celebrations of colonisation as well as scientific and technical modernisation in food production, the historical focus of shows has been on competition to maximise perceived quality and yield of goods, from wheat to cattle. Through these frameworks, shows are primarily represented as supporting industrial-scale agricultural practices that promote an ecologically blind approach to food production. However, by drawing on a case study including interviews with small-scale show exhibitors and judges in the horticultural produce categories and participant observation at a recent Royal Canberra Show and in the gardens of these exhibitors, I suggest that participation in the show can heighten placed-based engagements with the food-system and act as a catalyst for sensitising participants to the inescapability of our human/more-than-human relational togetherness (Turner, Henryks, Main & Wehner, 2017). This is particularly apparent when we focus attention on the role and formation of taste in these sites.

DOI: 10.4324/9780429424502-6

While show competition, at first glance, seems to move small-scale production beyond concerns with the pleasures of gustatorial taste—shown to play an important role in 'moving' participants in the gardens and AFNs explored in the previous chapters—to more mainstream industrial agriculture concerns with appearance, uniformity and quantity, the exhibitors and the judges we will meet are also intimately attuned to the productive and potentially disruptive more-than-human elements involved in food production. Perfect specimens are desired, but the people involved acknowledge that this is only achieved with an element of 'luck' where just the right relationships between human and more-than-humans congeal in the garden. And even then there's always something waiting to disrupt a gardener's or cook's best-laid plans. These are visceral, embodied experiences, attunement to which is shown to encourage practices marked by playful tinkering. While taste is identified as something that can be 'learned' and worked at, it is overwhelmingly seen as being produced through relational entanglements among often unknown and unknowable entities. In these ways, we see evidence of how a 'training of sensitivities' can enable the emergence of responsiveness to unexpected tastes in unexpected places formed through recognition of togetherness-in-relation.

To develop these arguments, I need to first introduce you to agricultural shows and position them within their historical context before familiarising you with the Royal Canberra Show and the tensions that become apparent in the staging of rural endeavours in urban locales. Once this background has been mapped out, we will meet the judges and exhibitors that contributed to the fieldwork and tease out the myriad ways in which engagements with taste can support a taste for togetherness at the Show.

Situating agricultural shows

Agricultural shows have been a focal point of my year for as long as I can remember. My maternal grandparents are life-long members of their local Show Society on the South Coast of NSW and my grandfather was regularly crowned best exhibitor in the horticultural section. His vegetable prowess saw him, surrounded by his garden, immortalised in a portrait on a bus shelter at the end of his street. The Show was integral to my grandparents' lives as they grew up with the Societies providing a social hub for many in their small farming communities. Throughout my childhood, this connection to the community remained important to them and each year they devoted countless hours as volunteers to help organise and run the major two-day event as well as preparing their own entries. Us grandchildren were enlisted too, helping take entry fees 'on the gate', being encouraged to exhibit and the girls who lived locally being hounded to enter the Miss Showgirl competition.

Miss Showgirl is an enduring feature of Australian Agricultural Shows where young women compete against each other to represent the local area at the larger regional and state-based shows. The girls are usually judged on their achievements, goals and knowledge of the agricultural industry. They also dress up in fancy clothes, wear sashes and get driven around the eventing arena in glamorous cars. My mother is one of four sisters (not the preferred gender balance for dairy farmers at the time) and while they all refused to be 'paraded around like cattle' at the show, in the last years of my grandfather's life my cousin Holly vied for the crown of Miss Showgirl. This sacrifice (and indeed it was!) is testament to the value my grandfather placed on the Show and the role he felt it played for the local community.

The significance of this local Show stretched out well-beyond the two days it was staged in early February. The Show Schedule, the document outlining all of the exhibition categories and the details of how the entries must be organised (for example: six pears, calyx intact), was the most thumbed book in my grandparents' home. My grandfather planned his yearly planting and harvesting with an eye towards having the produce hit its peak at Show time. In true farmer style, he was always circumspect about his chances. The 'conditions' and 'weather' were invariably not in his favour. Yet, every year, successful exhibition at the show remained his goal. Severely crippled with arthritis up until his death, he spent time everyday tinkering in his garden, getting ready for the show.

The first Australian Agricultural Show was held in Hobart in January 1822, shortly after the establishment of the Van Diemen's Land Agricultural Society on the 8 December 1821. In the following decades, agricultural societies were established across the Australian colonies with their shows often becoming one of the first community events organised in newly settled towns (Darian-Smith & Wills, 1999). The first agricultural show in NSW, 'the Parramatta Fair', was held in October 1824, with the precursor to the Sydney Royal Easter Show, the Metropolitan Intercolonial Exhibition, held in 1869. By 1893, the growing importance of Sydney's show resulted in its designation as a 'Royal' Show (Mant, 1972, p. 50). The Agricultural Society of NSW identified the Show's aim as being the 'encouragement and development of the primary resources of the state, chiefly through competitive shows' (Mant, 1972, p. 37).

In the early days, medals were awarded to pastoralists with the best 'sheep, cattle and boar, the best acre of wheat, barley, artificial grasses, largest crop of potatoes taken from three acres of land and for a collection of vegetables' (Mant, 1972, p. 17). Improving wheat yields was a key focus of the NSW Society, one which continued into the twentieth century, most evident in the field growing competitions that commenced in 1916 (Mant, 1972, p. 68). The shows' successes in 'popularising improved scientific methods of wheat farming and in introducing new disease-resistant varieties of wheat' prompted the 1926 introduction of maize and pasture and fodder field competitions. Beyond these competitive categories, the Society also performed other tasks including the gathering of climatic and disease data and fostering agricultural productivity (Mant, 1972, p. 37). The official focus was not on encouraging production of 'tasty' food but rather indicative of a 'taste' for the most productive foods that could support the nation both directly, through provision of supplies, and indirectly through the development of export markets.

Competition, colonialism and socio-technical progress

Academic analyses of agricultural shows in Australia tend to position these events as celebrations of the perceived civilising force of colonialism (Darian-Smith & Wills, 1999; Scott & Laurie, 2010) and of the socio-technical progress of modernity designed to promote a human/nature separation (Anderson, 2003). In these renderings, shows are depicted as staging the triumph of colonial man's dominance over non-humans, historically depicted as including Indigenous Australians through to less overtly lively things such as soil, that are forced to yield to the will of the white colonial male. The focus on progress also positions rural colonial man as the embodiment of modernity couched in terms of human exceptionalism typified through engagement in a myriad of socio-technical practices. As Anderson (2003, p. 423) writes 'the genre of the agricultural show enacts in thoroughly ritualistic fashion a triumphal narrative of human ingenuity over the non-human world'. Through shows and show societies, science and technology have been presented as the tools farmers could use to tame these nonhuman entities and reinforce the narrative of human separation from, and dominance of, the natural world (Anderson, 2003).

In the ninetieth and early twentieth centuries, show societies adopted the motto 'practice with science' (Anderson, 2003) and presented themselves as key to the nation's advancement and success, feeding its population and underpinning its export economy. This faith in science and technology as key to controlling the land and its potential harvest was a hallmark of agricultural practices in Australia and other colonial nations, often manifesting in an abstraction from place and, sometimes, contempt for local ecological conditions and knowledge. Official tastes and goals were decidedly industrial in flavour. This led to what Muir (2014), drawing on Arthur Crosby, calls 'ecological imperialism' whereby non-native species were grown by whatever means necessary. This drive to cultivate was fuelled not simply by a need to produce food (or excess food as Muir details) but by a social need to position agriculture as a means to 'civilise' settler nations (Muir, 2014, p. 4). Muir notes that '[t]he developing field of scientific agriculture could deliver a new class of technically educated, semi-professional workers and small landholders for the new century' (2014, p. 4).

In such approaches, the relationally formed materiality of food itself, including its taste, is understood to be controlled and manipulated by these skilful workers. While the colonial overtures of Australian agricultural shows may today have been dampened—as evident in the recognition of the hybrid natures of the produce and livestock on show which indicates at least partial adaptation of imported species to local conditions, breed and variety (Anderson, 2003)—the show classes still tend to be judged against generic criteria divorced from engagement with the conditions of particular localities found across the Australian landscape. Yet, as we shall see in the judging of entries in the small-scale horticultural categories at the Royal Canberra Show, there is evidence that, while judging criteria appear to be universally applied and static, the responses to, and enactment of, tastes in these arenas suggest a more contingent, relational approach.

Indeed, here we find that these encounters are understood as being formed through relations among biological response, material attributes of foodstuffs and social, economic and environmental conditions, all of which coalesce to induce humans to act in particular ways. These embodied responses to taste in competition at the Royal Canberra Show resist simple categorisation and narrativisation.

The spectacle of the Royal Canberra Show

Almost 200 years after Agricultural Shows began in Australia, competition has been found to still be the 'heart and soul of the show' (QCAS, 2012), but the public have demonstrated a 'taste' for more popular elements of the event. As nation-wide concerns about falling visitor numbers take hold, show societies have responded with an enhanced focus on the spectacle and entertainment aspects. While side-show alley—replete with rides, showbags and dagwood dogs—has been an integral feature of shows throughout the twentieth century, recent years have seen greater focus on spectacles in the eventing arena (beyond show jumping, parading animals and show girls). This shift has been attributed to perceptions of a growing divide and disconnection between urban and rural Australia (Henryks, Ecker, Turner, Denness & Zobel-Zubrzycka, 2016). At the 2014 Royal Agricultural Society of the Commonwealth Conference, Craig Davis, an international marketing expert, noted that in Australia:

> The vast majority of people today don't know a farmer. Fifty years ago, 14% of the population were involved in agriculture, that figure is now 0.6%. There is a huge disconnect amongst an urban centric society.
>
> (Davis, 2014)

This lack of direct connection, fuelled by distanciation between places of food production and cities where the majority of the population reside (Blecha & Leitner, 2014; Steel, 2009) has been both identified as a barrier to urban dwellers' involvement in shows as well as a marketing opportunity enabling an emphasis on the spectacle of rural 'otherness' on display at these events (Henryks et al., 2016). This is evident in the strategies employed by the Royal Canberra Show which aims to provide a site for 'city meeting country' and 'country meeting city' (Royal Canberra Show, 2014).

The spectacle of rural otherness on display at agricultural shows has been identified as an attempt to attract city dwellers by presenting a version of rural Australia that aligns with their imaginaries (see Scott & Laurie, 2010). This has become particularly apparent since the 1970s with the addition of growing numbers of exhibitions and events that serve to provide more accessible and palatable engagements with rural Australia. This includes the introduction of animal nurseries, events and displays that depict 'a romanticised rural past' and the staging of rock concerts to attract more visitors (Scott & Laurie, 2010, p. 35). These have all occurred at the Royal Canberra Show with a particular focus in recent years on attempts to appeal to the urban, young adult demographic. In

2014, the president of the Royal National Capital Agricultural Society (RNCAS), Stephen Beer, noted 'We were aiming for the 18 to 25-year-old market with our special attractions on the main arena' (Kretowicz, 2014), which included free-style motocross and Smash Up Demolition derby. However, the inclusion of greater diversity of entertainment at shows is not seen to eclipse the rural endeavours and competition. Indeed, the QCAS report found that competition remains a key 'community engagement strategy' (2012) and Stephen Beer asserted that the 2014 Canberra Show 'was definitely an agricultural show, with an educational slant' (Kretowicz, 2014).

And so, we now turn our attention back to competition, the 'heart and soul' of the show, with the remainder of this chapter focusing on the small-scale producer categories at the Royal Canberra Show, the way taste is configured here and how this impacts on the producers and judges. To do this, I analyse the multi-layered engagements with place and more-than-humans identified in conversations with, and observations of, eight exhibitors and three judges in competitive horticultural classes from fresh fruit and vegetables to preserves and cakes. Here, we will encounter exhibitors who, through ongoing playful tinkering, enact and are enacted by, relational modes of togetherness through food production. These producers are intimately aware of the limits of their control (even when control continues to be the aim of their practices) and adept at living with uncertainty. It is through the contingent characteristics of these encounters, and responsiveness to the multifaceted configurations of taste, that I identify the potential for different ways of being and doing in the world that can be fostered through practices of playful tinkering enacted with convivial dignity. To explore these possibilities, we begin by turning to the role of taste in judging to see how the relational materialities of food production coalesce in the taste of foods and judges alike.

Taste, judging and being moved at the show

At the 2012 Royal Canberra Show fruit and vegetables 'staged' on plastic plates in the horticultural pavilion were rapidly decaying in the humidity of late summer that had induced severe storms across the region in the previous weeks, hampering the gardening efforts of the eight exhibitors followed in this research. By the show's end, the fruit and vegetable entries, particularly those sliced open for judging, had taken on distorted forms, collapsing in the heat and some barely recognisable under the blanket of feasting fruit flies and encroaching mould. Ninety-eight thousand people attended the show with the horticultural pavilion becoming a site-in-flux, 'co-constituted and performed by human and nonhuman actants alike' (Cloke & Jones, 2001, p. 655). Seats were scarce to observe the Thursday morning judging of the jams, spreads and preserves and cookery sections. The judging was slow and steady with key points about the entries announced loudly so the audience could jot down notes with hopes of improving their chances of winning next year. Accompanied only by a steward (and sometimes the researchers on this project), the horticultural judge acted without an audience, moving purposively along the produce tables inspecting and making notes before prize certificates were written out and set in place (Henryks et al., 2016).

Unlike the jams, spreads and preserves, the fresh fruits and vegetables entered into the Royal Canberra Show are rarely tasted by judges. Gustatory taste does not always figure in the judging guidelines. As we saw above, agricultural shows from the get-go have been focused on industrial agriculture practices with a particular focus on improving yield. Within these industrial-scale frameworks, quality is largely judged by appearance, size and form, with a notable interest in uniformity. These are characteristics that dominate global food flows, leading to farm gate rejections of crops and homogeneity in supermarkets. The challenges encountered in attempting to 'grow to Schedule' (the Show Schedule being the document that sets out the judging criteria) to meet these largely cosmetic guidelines that are, apparently, characteristics of the 'best' natural specimens (as such, the way they should look and be—a particular version of pure nature—if other nonhuman elements are kept away from them and humans provide the correct care to enable this) serve to highlight the complexities of production while also motivating the gardeners' efforts to win categories.

Yet, this focus on the appearance of the fruit and vegetables at the show is not necessarily at the expense of their taste, as many of us have experienced with much of the fresh produce sold in mainstream supermarkets. As explored in the previous chapters, the affective force of pleasurable gustatory taste is a key motivator for small-scale food-producing gardeners. Yet, this is often not critical to winning the competition. Instead, visual appeal dominates at the show. This is taste of a different kind, a taste for a particular aesthetic form where appearance can cause moments of arrest, immediately 'catching the eye'. At the 2012 Royal Canberra Show, the winners of the horticultural produce were quite obvious to the judges as they caused them to be immediately moved by the foods. One judge, upon completing his assessment of the junior entries in a 'collection of vegetables' noted that the first place was awarded to a specific collection:

> . . .because of its uniformity, size is good, freshness. It includes a large button squash, beetroot and some potatoes and some peppers and capsicums and some black Russian tomatoes and some well-presented beans and three cobs of corn. And it's not a lot. . . not a great lot of difference between the others, but it's uniformity, appearance, it takes your eye as soon as you see it and that's why I awarded it the best junior collection.

The visceral qualities of the foods, the appearance of freshness evident in their vibrant colour conjuring up the snap of the beans or the crunch of the capsicums together with the uniformity of each vegetable type and, as he noted later in our discussion, the 'amount of varieties' in the display induced an immediate response. The visceral here, as the Hayes-Conroys' (2013) contend, is an embodied experience that includes, rather than excludes, the mind. The judge was trained through his previous sensorial experiences to have a taste for certain characteristics of the fruits and vegetables but it is the effect of the collection together that 'takes the eye' and leads to the awarding of the prize.

Temporal, contingent and non-narratisable: attuning to the visceralities of taste

Conversations with the judge revealed his acknowledgement of the contingent nature of food production and the aesthetic tastes applied to these. Speaking again of the winning junior collection he observed 'they come up really good on the day, and that's why I chose that collection'. Here, he indicates his attunement to the temporal qualities of producing prize-winning goods. Things could easily have been different on another day. The aesthetic 'taste' of the judging involves drawing connections to the challenges of production encountered by the exhibitors and acknowledgement of the desired and necessary detachments (from potentially damaging weather, pests and diseases, etc.) needed to win. Other entries facing similar challenges as well as affordances had comparable qualities but the winner had something else that, while difficult to articulate, was readily perceived by this judge.

This may well be representative of what Carolan calls the 'more-than-we-can-tell practices' (2016, p. 147) of food, those things which exceed language. He refers to these as constituting '"sticky" knowledge' because it 'doesn't travel well' (Carolan, 2015, p. 128). Such sticky knowledges are regularly encountered when attempting to convey taste, but here where other senses (primarily the visual) are called upon it is also evident. The judge admitted he found it challenging to convey the reasoning for his decisions in words. The discussions were being recorded for an online exhibition (including audio) to be hosted by the National Museum of Australia and so, before we began our chats, our research team had asked the interviewees to try and repeat our questions in their responses to assist with the later editing of the audio files. After we recorded the above comments the horticultural judge stated:

JUDGE: Not quite what you wanted was it?
INTERVIEWER: No, no, that's good. That's good. So we might...
JUDGE: I'm not a speaker I'll admit.

His response to the exhibits was embodied, immediate and visceral. It was also relational. While he could point to some particular aspects of the judging criteria that the winning exhibits had excelled in, notably uniformity, overwhelmingly his judgment responded to the relations among the various judging criteria and the exhibit itself that produced the overall winner. But the relational entanglements that induced the privileging of these particular tastes were difficult to make sense of in words.

These relational entanglements were brought to the fore in the entries the judges didn't just assess visually but those they also touched, smelled, sliced, peeled and/or tasted. This multisensorial engagement is enacted to assess conformity to 'ideal types' and correct 'staging'. For example, this can include, for a tomato, making sure the stem has been removed. Apples and pears on the other hand must have the stalk and calyx intact. Rhubarb must not be cut from the plant but torn, and its leaves need to be trimmed. Potatoes must be brushed, not washed. While these modes of presentation are written out in the show schedule they are also embodied, visceral acts that impact on engagement with the foods and, in some categories, allow certain qualities of the broader plant, tree or soil to impact on the judging. It is the moment of flesh being cut or husks being pulled back that determines the internal 'quality' of the food and whether it aligns with the relational tastes of the judges. Just as play involves entering into risky, unknown relations, here the propositional nature of these relations are open to assessment. Internally, tomatoes could be malformed or corn kernels could be desiccated. This was an incredibly nerve-wracking time for our exhibitors, with one observing:

> Corn is the unknown quantity because you don't know until the judge peels back the husk as to what quality of corn you have, what damage the pests have done etcetera, etcetera. So it's very much an unknown quantity. But if you can keep the water up to them then generally speaking you can get a good group. I don't think my corn is as nice as some of the other exhibits here, in fact I think I might be lucky to get a second.

This not knowing, the provision of human care and the 'feel' of the unveiling point to the ways growers and judges are attuned to the embodied, viscerality of relational becomings-with food. These attunements, as we can see in the corn example above, can afford both attachments (to the desired production of prize-winning juicy corn) and detachments (pests). Attunement in gardens and in production for show exhibiting does not lead to knowing. Instead, these processes largely manifest in recognition of, and a sense of respect for, the alterity of others and the unknown and unknowable outcomes of the relational entanglements of becoming-with. This leads to a responsive form of togetherness marked by recognition of difference but acceptance of the necessity of living together which manifests in a form of togetherness-in-relation that I refer to as being guided by convivial dignity. Gustatory and aesthetic taste judgments draw attention to the uncontrollable generative potential of these relations.

While the foods exhibited in the show are judged according to notions of 'perfection' where industrial agricultural qualities such as uniformity are key, this does not lead to seeing humans as being in control nor one of viewing the plants as simply performing a natural, innate role. Instead, as Mol, Moser & Pols (2010) have found to be the case with industrial-scale tomato growers and which I observed occurred in the practices of the gardeners discussed in chapter three, these food producers tinker with production methods to grow ideal types to win at the show. However, at the small-scale production level, this tinkering is also marked by playful engagement. As we have seen, play is characterised by encounters among unknown and unknowable entities that, through their interactions, can generate something new or unexpected. Play can enable productive experiences of togetherness that draw on and produce both attachments and detachments, but which consistently highlight the process of becoming-with as necessary togetherness.

Growing for the show involves playful tinkering practices often motivated by gustatorial and aesthetic tastes that highlight mutual vulnerabilities and draws attention to the limits of all entities involved. The forms of play glimpsed here induce encounters with uncertainty that are emphasised due to the temporal and spatial necessity of growing for judging day and the tastes of judges. Play, when enacted with convivial dignity, further highlights the necessity of mutual responsiveness and the limits of human control. It emphasises the inescapability of the togetherness of humans and more-than-humans but provides a means of conceptualising this without privileging the need for intimate familiarity through conceptions of care or attachment. Exhibiting at the show is a risky, playful proposition of togetherness-in-relation.

Tinkering with uncertainty and attuning to the unknowable: playing with a taste for togetherness

The focus on being 'best in show' by producing perfect food specimens not only serves to highlight a taste for the privileging of a particular aesthetic that seems to reinforce values embedded in the industrial food system but has the potential to reinforce desires for detachment from more-than-human others that could thwart ones' chances of winning. When discussing the specific practices carried out in their gardens in the lead up to the show, the exhibitors spoke of the difficulties faced in relation to weather and pests and the impact these had on their capacity to harvest uniform produce. Most exhibition categories require more than one specimen to be entered. As such, what is being judged on the day is not the capacity to produce and care for one 'ideal type' but, as is reflective of the show's historical focus on large-scale agriculture, the ability to replicate this. Categories may require three tomatoes or six beans and these need to be as similar to each other, and the specifications set out in the show schedule, as possible. To flesh this out a little, I am going to step away from plants momentarily to dwell with eggs and the story of one husband and wife (Chris and Merran) team who exhibits them at the Royal Canberra Show. Here, while taste may not appear to be front and centre in the discussion, the relational enactments of gustatorial and aesthetic taste motivates the analysis and enables identification of the tensions and generative potential that recognition of togetherness-in-relation can induce when growing to exhibit in the show.

When entering eggs into the show, a dozen must be presented and, as freshness also forms a tacit part of the assessment, the eggs need to be laid close to show time. The challenges this induces are outlined by Chris:

> I entered three dozen eggs this year. Two lots were brown and one of white. The idea is over a period of a couple of weeks you try and get all the eggs to be the same colour, shape and size, so that's how we did our white ones. The same with our brown ones. The chooks aren't always cooperating. And we stage our eggs with the pointy side down and that's how we store our eggs at home. We don't store our eggs in the fridge because we don't have a rooster so they're all non-productive.

The eggs at the Royal Canberra Show in 2012 were judged by the horticultural judge and were assessed on their external appearance. This was something Chris, who received a First Place for his white and second for his browns, was conscious of as he spoke about the 'staging' of the exhibits and the need to clean the eggs when collected:

> We've found that if the eggs are presented and they're clean and there's no mud marks on them, if you leave mud marks on the eggs for too long they do physically stain the egg and you just can't clean it off. So as you collect the eggs on a daily basis you then have to clean them straight away otherwise, as I say, that stain will occur.

Here we can see that the knowledge and practices at play in producing prize-winning eggs are developed in a tinkering, adjusting mode. However, this was not simply reliant on the human capacity to learn and adapt but was also spoken about as a multi-species partnership, one in which the human capacity to impose their will is limited, as evident in the recognition that 'the chooks aren't always cooperating' in the human endeavour to produce foods that fit the aesthetic tastes of the judges. Attunement to the chickens does not produce knowability and most certainly does not enable control of what is produced.

In our conversations, Chris and Merran constantly shifted between statements expressing desire for control in the garden—over produce and animals—while also describing attunement to the relational entanglements that produced these foods and to which they were responsive even if, at times, they found this extremely frustrating. As Chris talked about getting together his egg entries for the show he observed:

> So, really it's down to the relationship between yourself and the chooks, if you start feeding them good nutritional food at least three or four weeks beforehand to make sure they get in good production. If you don't feed the chooks correctly they don't produce, like a lot of home animals, pets etcetera, if you don't look after them and make them fit and healthy then you won't get the product at the end.

In this 'talk', the relationship between chook and carer appears to be one-way—privileging the tinkerer/carer but, when later expanding on the ways in which

Chris and his wife tended to their garden and 'lived' together with the chooks, the relationships were revealed to be far more nuanced. The chickens were an integral part of garden management and entities that they lived together with. Sometimes this involved attachments and at other times detachments but they were consistently relationships that involved recognition of togetherness-in-relation in a way I conceptualise throughout this book as being representative of the alternative grammar of convivial dignity.

The bodies of the chickens, while prompting some desire for detachment (from the 'poo', euphemistically referred to as mud, on the egg, which is simply matter out of place here as it is a valued addition to the garden to improve fertility when found in other locations) also assisted the exhibitors to 'detach' from more harmful or disliked others (slugs and snails that threatened their potential fruit and vegetable entries in the show). These detachments were aided by observation that enabled attunement to the garden and its multispecies inhabitants in ways marked by tinkering play, insofar as they were experienced as mutually responsive, generative relations. Speaking of engaging with their grandchildren in the garden, Chris noted:

> Observation is a major thing. Over the life, those 12 weeks of plants growing to maturity, they see all the stages of it and they're [the grandchildren] active in cursing the snails if they've eaten a couple of their plants. They've put stuff in our garden as well as their own.

With Merran adding:

> They collect the snails and take them to the chooks, they feed the chooks. The children actually collect the snails and take them to the chooks, the chooks have a go.

Here, attunement and togetherness is shown to be able to induce attachment and detachment. The relations are not always guided by an ethic of care. But it is not these influx relations that are of concern here because, despite the mode of relation and its affects, these exhibitors and their broader family recognise the inescapability of responsive forms of togetherness-in-relation. They cannot act alone here, and all of their actions are enabled by myriad other relations. As I contend occurs in play guided by convivial dignity, these were bodies that were not only moved by others but which could rarely move others in an intentional premeditated fashion. Thus, this form of togetherness meant learning to live with the knowledge and experience that it was never possible to know or predict what the outcomes of relational interactions would be. The modes of togetherness were in a constant state of play and required engagement in practices guided by the ethico-political conception of convivial dignity.

The eggs produced through these relationships of togetherness outlined above went on to win prizes at the Show. They immediately stood out as winners to the judge. Indeed, as we transitioned from speaking about the vegetable collections to eggs, the judge noted of the brown entries:

JUDGE: There are some good eggs here. That... those dozen eggs around here got the best exhibit.
INTERVIEWER: They're beauties... They stand out.
JUDGE: You can see they're a stand out, there's no doubt. Do you want to have a look at them?
INTERVIEWER: Yeah.
JUDGE: I wonder if I can...
INTERVIEWER: The white ones or brown?
JUDGE: The browns. They're beauties.
INTERVIEWER: Yeah, they look perfect, don't they?
JUDGE: Look at that.
INTERVIEWER: Yeah. Fill up the cup.
JUDGE: Look at that. You can just see...
INTERVIEWER: Yeah. The presentation.
JUDGE: ... right, it's a bit of a difference.
INTERVIEWER: Yeah.
JUDGE: There's no comparison. Just look at that. They're just... they've got the best exhibit.

Again, the precise qualities that made these eggs 'stand-out' could not be clearly articulated in words but it was immediately obvious to the judge. As he pulled them down, carefully extracting one and turning it gently in his hand, he remarked again, 'they're beauties'. Here the eggs conform to the desired aesthetic taste, they are capable of being identified as 'beauties' and the unspeakable visceral, embodied recognition of this prompts responsiveness, ensuring they won best exhibit. However, not all categories in the horticultural pavilion are judged according to aesthetic standards alone. For some, engagement with gustatory taste is critical.

Relational tastes on show

Gustatory taste plays a key role in the judging of the jams and preserves section at the Show. Here, relational conceptions of taste are expressed through the entanglements of appearance, consistency (mouth-feel) as well as flavour. In observations of judging and conversations with the head judge at the 2012 Canberra Show, flavour was the most remarked upon quality in the judging of pickles and preserves. The capacity to discern this gustatory pleasure enveloped within the 'taste' of the produce was not discussed as an inherent, natural skill but something the judges had been 'trained' to become attuned to. The desired tastes were, however, not seen to be static nor universal. Other judges were presumed to be responsive to different tastes, thus capable of detecting and rewarding other qualities. Again, as we saw with the horticultural judge above, despite the written criteria against which entries were judged, the capacity to articulate the stand-out characteristics of the winners was something judges struggled with. As one judge observed, it was all about the capacity to 'get the flavour':

> The one who the... zucchini pickles that won the champion today, it was zucchini and mustard and it had a beautiful flavour. You could taste the mustard, taste the zucchinis where a lot of the others were so spicy you can't get the flavour. But it was beautiful and I sort of picked it straight away for... yeah...
>
> ...because it's mustard, you need to be able to taste the mustard, and the zucchini went well with it. So as soon as I tasted it I knew it was a winner. The others... a lot of the others are too spicy for me, other judges might have different ideas. But my taste was for that one.

Just as we saw in the judging of the collection of vegetables and eggs, this mustard pickle immediately moved the judge. The visceral, embodied engagement with this matter produced a taste—a flavour—that motivated responsive action. This was a product of the relations among the pickle's ingredients and the learnt 'tastes' of the judge. It wasn't any one particular ingredient that made this a winner but the ability to detect a variety of flavours of the various components worked together to produce a prize-winning taste.

The 'training' or 'learning' of taste through the multisensorial and material components of judging is displayed to exhibitors through the performance of the judging of the jams, jellies and preserves. Here, eager exhibitors gather to observe the judging process to attune their tastes to the expectations of the show judges. This was exemplified by one of our entrants, Cheryl, in the jams, jellies and preserves section for whom the judging is an annual social occasion she attends with two of her close friends:

> for the last two or three years, I've actually been going to the judging, which is by far the best way to learn how to make [and] what you're doing wrong. And you can, you, because the judges usually talk while they're looking at everything and holding things up and telling people what's the matter. So, it's by far the best way to hear what the issues are...
>
> My understanding is the shows change the judge every year, so you can't cook for a judge, because people will. I know I would. So, yeah. You've always, never sure exactly what the judge is looking for. So, there's no point cooking for a judge.

For this exhibitor, taste can be learned to fit in with the expectations of the show. It does not manifest as an innate quality of the foods themselves but as the taste desired in this setting. While the judge of the zucchini pickle above links this to individual tastes, for this exhibitor, taste was more broadly linked to what she saw as being the Show's preference for 'old-fashioned' tastes.

Judging day was a time of nervousness, excitement and learning for Melissa, 2012's champion pickle exhibitor. As she notes:

> the judging is interesting to watch. I sat in and watched several. They picked the different ones that they have a taste test first and it must be quite difficult

> because some of the categories are all one variety and the ones that are, say, savoury sauce can be all sorts of different ones. So, you have to have a real talent in knowing what you're doing to be able to pick them, one sauce over another, when they're in different types of sauces. And I suppose doing the mustard pickle, I looked at mine and looked at the colour of the others and I thought I haven't got enough turmeric in it, it's too white. And then when she pulled it out I thought 'Oh, my goodness, is she going to give me a prize here?', but I didn't know whether I was first, second or third. And then she saw my excitement, I think, when she pulled it and she said 'Is one of those yours?' 'Yes,' I said...
>
> My other sauces didn't do so well, but I guess that's going back to understanding what you need to do in the categories 'cause I hadn't entered before. I was a bit unsure about tomato sauce, is it tomato sauce you're going to put on sausages or is it tomato sauce for pasta or there's not really a clear line. So anyway, I questioned it when I brought it in and put my sauces in different categories, one of them anyway, and... but still wasn't good enough.
>
> But I'm learning, so I'm learning some new tricks, sort of they should be pourable and some of them weren't today, and we saw the different varieties and sauces that were a lot thicker and you can see which ones were chosen, getting some ideas of the flavours that were required to win.

Taste in the jams, pickles and preserve encompasses appearance, flavour and presentation and was experienced through a process of attuning to a variety of expectations across numerous characteristics. As we see in the judges' comments, this learning is not just about formal knowledge whereby rules are followed to the letter. Instead, it is largely a matter of embodied practice. It is only in the process of the embodied, relational 'doing' of taste that a winner can be discerned. Bodily judgments and the multiplicity of material, social and economic relations that congeal in this act in 'more-than-we-can-tell' practices (Carolan 2011, p. 60) are integral here. While difficult to explain in words, these relational encounters are keenly sensed, producing immediate visceral responses particularly evident in taste. The winning pickle appeared too 'white' to the exhibitor but the judge's immediate response to its flavour overrode any issues with its constituent parts. The relational entanglements that produced its taste enabled the particular elements to be exceeded, to become more than the sum of its parts.

Conclusion

Agricultural shows, through their support of industrial scale food production, have historically been positioned as perpetuators of the civilising forces of colonialism (Darian-Smith & Wills, 1999; Scott & Laurie, 2010) and of the socio-technical progress of modernity. However, in this chapter I have suggested that an attentive approach to the practices of small-scale exhibitors and those judging these categories can elicit a more nuanced reading where participation in shows is demonstrated to be capable of offering opportunities to rethink and redo anthropocentric norms of human exceptionalism. This becomes particularly evident through exhibitors' attunement to the relational entanglements of taste, in both its gustatorial and aesthetic guises, and their attempts to enter products that conform to the tastes of judges.

In preparation for the Show, small-scale producer encounters with the limits of human control and the necessity of engaging in relations with unknown and unknowable others where the outcomes cannot be predetermined are shown to manifest in recognition of the inescapability of togetherness-in-relation. While this is often a source of frustration, its inevitability means the producers regularly develop ways of living together with more-than-humans that are responsive to this uncertainty. The judges too are confronted with the relationalities of taste on judging day. Here, the recognition of togetherness-in-relation is shown to be unable to be expressed through existing narratives and grammars that tend to reproduce anthropocentric norms. Instead, it manifests in visceral, embodied experiences. This highlights the necessity of the development of alternative grammars able to speak to and amplify ontologies that support entangled ways of being and doing that promote more sustainable living behaviours. While attention to the materialities of togetherness are identified as key to this, I also point to the necessity of semiotic responses that speak back to dominant narratives of human exceptionalism. This material-semiotic approach can unfold in a playful tinkering approach through attunement to taste when enacted with the alternative ethico-political grammar of convivial dignity.

This chapter marks the end of our journey with taste and the beginning of our explorations with waste. With this new beginning comes the opportunity to consolidate this material-semiotic resistance to anthropocentric practices and beliefs. To do this, I stay close to matter and the relations that enliven it to explore the generative potential of visceral, embodied food waste encounters. There, in the murkier world of death and decay, we will see that waste, like taste, is induced by shifting relational entanglements that escape human control. Attunement to these encounters with human limits and experiences of togetherness-in-relation are shown to be capable of promoting agile, creative, resilient modes of being and doing in the world that exceed the limits of anthropocentric imaginings.

However, these ontologies and potential world-making practices are once again shown to resist narrativisation. This results from their in flux materialities as well as the limitations of existing grammars that tend to focus on evoking conceptions of care and attachments to support uptake of more sustainable ways of living. Drawing on the fieldwork explored in the following chapters, I will once again argue that interactions marked by playful tinkering guided by convivial dignity may offer one approach to destabilising, and representing, alternative modes of being and doing with the world. To lay the foundations for these arguments, the following chapter maps out why attention to food waste is so critical in our contemporary context by identifying the generative potential of reconfiguring social, environmental, economic and material food waste relations.

References

Anderson, K. (2003). White natures: Sydney's Royal Agricultural Show in post-humanist perspective. *Transactions of the Institute of British Geographers*, *28*(4), 422–441.

Blecha, J., & Leitner, H. (2014). Reimagining the food system, the economy, and urban life: New urban chicken-keepers in US cities. *Urban Geography*, *35*(1), 86–108.

Carolan, M. (2011). *Embodied food politics*. Burlington, VT: Ashgate Publishing Ltd.

Carolan, M. (2015). Re-wilding food systems: Re-wilding food systems: Visceralities, utopias, pragmatism, and practice. In P. Stock, M. Carolan, & C. Rosin (Eds.), *Food utopias: An invitation to a food dialogue* (pp. 126–139). New York & London: Routledge.

Carolan, M. (2016). Adventurous food futures: Knowing about alternatives is not enough, we need to feel them. *Agriculture and Human Values*, *33*(1), 141–152.

Cloke, P., & Jones, C. (2001). Dwelling, place, and landscape: An orchard in Somerset. *Environment and Planning A*, *33*(4), 649–666.

Darian-Smith, K., & Wills, S. (1999). *Agricultural shows in Australia: A survey*. Melbourne, Vic.: The Australian Centre, University of Melbourne

Davis, C. (2014). *Agriculture: The greatest show on earth*. Paper presented at the Sustainability of Tomorrow's Agricultural Shows, Brisbane. Retrieved from http://www.therasc.com/rasc-2014-conference-programme/

Hayes-Conroy, J., & Hayes-Conroy, A. (2013). Veggies and visceralities: A political ecology of food and feeling. *Emotion, Space and Society*, *6*, 81–90.

Henryks, J., Ecker, S., Turner, B., Denness, B., & Zobel-Zubrzycka, H. (2016). Agricultural show awards: A brief exploration of their role marketing food products. *Journal of International Food & Agribusiness Marketing*, *28*(4), 315–329.

Kretowicz, E. (2014, February 23). Another happy crowd of 100,000 enjoy Royal Canberra Show. *The Canberra Times*. Retrieved from http://www.canberratimes.com.au/act-news/another-happy-crowd-of-100000-enjoy-royal-canberra-show-20140222-339cg.html

Mant, G. (1972). *The big show: 150th anniversary Royal Agricultural society of New South Wales*. Sydney: Horowitz Publications.

Mol, A., Moser, I., & Pols, J. (Eds.). (2010). *Care in practice: On tinkering in clinics, homes and farms*. Bielefeld, Germany: Transcript Verlag.

Muir, C. (2014). *The broken promise of agricultural progress: An environmental history*. London & New York: Routledge.

QCAS. (2012). *An economic & social impact study of Australian agricultural shows*. Brisbane. Retrieved from http://www.queenslandshows.com.au/pdf/FINAL-Impact-Study-Report-190612.pdf

Royal Canberra Show. (2014). *About the show*. Retrieved from http://www.canberrashow.org.au/about-show

Scott, J., & Laurie, R. (2010). When the country comes to town: Encounters at a country Metropolitan Show. *History Australia*, *7*(2), 35-1–35-22.

SRES. (2014). Media release: Learning is fun at the 2014 Sydeny Royal Easter Show [Press release].

Steel, C. (2009). *Hungry city: How food shapes our lives*. London: Vintage.

Turner, B., Henryks, J., Main, G., & Wehner, K. (2017). Tinkering at the limits: Agricultural shows, small-scale producers and ecological connections. *Australian Geographer*, *48*(2), 185–202.

7 Introducing waste

Introduction

So many people I talked to in the course of the research for this book emphatically stated 'I hate waste'. The production of waste induced powerful affective responses prompting many to invest significant amounts of effort to reduce it. For others, it induced guilt, worry and anxiety about their excess and the manner in which it seemed to escape their control. Through these stories and experiences, I was reminded of my amusement the first time I saw my grandmother wash out plastic bread bags, pegging them on the clothesline to dry ready for repurposing. There were dozens of them, a result of my grandfather's regular collection of 'old' bread discarded by supermarkets and bakeries because it was no longer deemed acceptable to sell. The bread was destined for his beloved pigs, but we always got first pickings. I developed a taste for many different loaves thanks to this waste. Through these material encounters, the instability of waste and its potential to be otherwise were conspicuously manifest.

In this section of the book, we shift our attention from the seemingly more pleasurable arena of taste to that of waste. While the focus of our food-related behaviours changes here, I aim to extend and consolidate my interest in paying close-attention to our multisensorial, visceral engagements with food. In so doing, as we have seen with taste, we will find that waste is also produced through shifting relational encounters. Waste isn't something that just is. It is produced through ongoing processes of becomings-with, attunement to which can support recognition of the relational entanglements of human and more-than-human lives and an openness to being moved by these encounters. However, as will become apparent, the fieldwork participants are found to not simply be moved by matter but also by the propositional and generative potential of the relations that enliven it. These responsive experiences prompt the development, understanding and appreciation of the necessity of our togetherness-in-relation.

As occurred in the chapters coalescing around taste, throughout the waste section, I will suggest that existing narratives that attempt to support more environmentally sustainable living habits draw on limited understandings of the relations necessary to motivate a reconfiguration of the dominant hyper-separated

DOI: 10.4324/9780429424502-7

notion of the modern human subject. This is a reconfiguration that I identify as being vital to underpinning the development of less resource intensive lifestyles that work towards ensuring a survivable future for humans and more-than-humans. As such, I contend that the task of shifting anthropocentric behaviours and beliefs requires material-semiotic work. While I call for greater attentiveness to relations of materiality, I also argue that we need to develop alternative narrative forms, or grammars, capable of representing and amplifying non-anthropocentric ways of being and doing in the world that attend to our future uncertainty in the present. Play enacted with the ethico-political guide of convivial dignity, and often unfolding in a tinkering mode, is further developed throughout the following chapters as one approach that could contribute to realising this.

To establish a solid grounding for these arguments, this introductory chapter on waste provides an overview of contemporary understandings of food flows through households, identifying the principal problematic aspects of waste insofar as the term appears to denote a static, inevitable material form that obscures its relational becomings. It then provides a brief overview of the global food waste problem, paying particular attention to its social, environmental and political-economic manifestations in policy and academic literature. To get us started on this journey, the following section highlights the generative potential of a rethinking of waste relations for encouraging non-anthropocentric modes of being and doing by drawing attention to the unavoidability of householders' engagements with waste's very vitalities and affective force.

Confronting excess: the generative potential of encounters with waste's vitalities

In Chapter 1 we saw that as eaters we must enter into a 'conversation with those who are not "us"' (Haraway, 2008a, p. 174). Attunement to the pleasures, dislikes and affective force of gustatorial taste experiences explored in the preceding section requires us, as Haraway writes, 'to strike up a coherent conversation where humans are not the measure of all things and where no one claims unmediated access to anyone else' (Haraway, 2008a, p. 174). These conversations are predicated on 'connections' that Gibson-Graham point out rely on the understanding that '[w]e are all just different collections of the same stuff—bacteria, heavy metals, atoms, matter-energy—not separate kinds of being susceptible to ranking' (Gibson-Graham, 2011, p. 3). However, the internal entanglements of eaters are well hidden beneath our human forms and food most often simply disappears within, sometimes marking our bodies in particular ways but becoming visible only through its human manifestation and the waste expelled by our bodies. The latter, thanks to modern sanitation in the minority world, is able to be readily removed to protect our health and that of our communities. Consequently, the human gift-wrapped multispecies entanglements that enact digestion effectively remain hidden from view. So, while eating and its co-conspirator digestion are key sites where relational entanglements and the vulnerabilities of the compositional nature of human bodies are evident, the very obfuscation of these processes and interconnections may well limit their capacity to prompt more relational, attuned approaches to human/more-than-human togetherness.

But the food that eaters don't eat—that which is left over because we cooked too much, or we bought too much and it went off, or which we decided we didn't like or just didn't feel like anymore, doesn't disappear in quite the same way. Excess food, its prevention, reuse and disposal, requires management through intimate human bodily engagements where the foods' very vitality is inescapable. Through the affective force of these engagements, which can prompt desires for both attachment and detachment and manifest in a myriad of forms of togetherness, mutual vulnerabilities in living together are exposed. In this section of the book, we will explore the potential for multisensory experiences with this excess, surplus or wasted food to provide a jolt for encouraging attuned, responsive forms of human/more-than-human engagement that could lay the foundations for less resource-intensive lifestyles attuned to togetherness-in-relation.

In the majority world, most bodies have been habitualised into routines that obscure waste. Cleanliness requires removal of waste from our homes through sewage systems or through the use of bins that provide a series of conduits connected to municipal waste collection services, many of which are increasingly managed by private providers. Discourses of health and hygiene alert us to the negativity and danger waste presents to our bodies (germs, maggots and mould are waiting to invade, threatening the health and bodily integrity of us and those we care for). Food waste presents as a particularly concerning hazard. It is messy; it rots, feeds bacteria and can emit unpleasant odours. Most poignantly, it can make us sick. To prevent this occurring, management of surplus food tends to require us to be 'hands on' rather than having the relative luxury of being at arm's length as we most often are when vacuuming, dusting or scrubbing the toilet. Commonly, this is a case of sorting through the fruit bowl and a regular cleanout of the crisper and fridge with the what-was-once-food being disposed of and the traces it leaves behind (both visible—such as mould, and invisible to the naked eye—such as the potential pathogens that could accelerate decomposition of nearby food and threaten our health) being eliminated by wiping out, washing up or some combination of the two that aims to restore the appropriate level of sanitation to the storage receptacle.

The intimacy and intensity of a necessarily 'hands on' approach to food management is heightened if the surplus is separated from other rubbish into a compost bin, bokashi bucket, to feed worms/chickens/pets or to send to an organic waste collection service. Where a number of these forms of food redistribution are in play, these processes can require more time and thought as people decide which waste is best for which redistribution channel, often developing an informal hierarchy of waste disposal. Such practices of waste redistribution can be more labour intensive than simply discarding waste into mainstream bins. Guides to worm farming and compost often recommend chopping up waste into smaller pieces to facilitate faster ingestion by the worms and decomposition by the microorganisms that we rely on to manage the waste. There are also some forms of waste that people are advised to avoid diverting to these management systems (for example no onions and citrus for worms, no avocado for chickens and no meat for compost). Thus, management of food waste and its very vitalities, from regular fridge and fruit bowl inspections through to the decision to discard, requires a significant amount of human labour. The seasonal fluctuations in our need to attend to this, with the creep of summer heat and high humidity catalysing ripening and decomposition, means that effective food management routines can rarely be predetermined. Instead, they must be responsive to climate, seasons and the characteristics of the food brought into the household. These attuned processes are both induced, and motivated by, a multiplicity of affects.

Food waste and our management of it are multi-sensorial, visceral experiences. Despite the apparent profligacy and associations with a 'throwaway society' mentality that are commonly connected to the large quantity of food waste produced in the minority world, the discarding of surplus or excess food waste tends to be accompanied by feelings of guilt (Baker, Fear & Denniss, 2009; Evans, 2012, 2014). The first large-scale research into food waste in Australia found that 84% of people reported feeling guilty when they throw away food (Baker et al., 2009; Evans, 2012, 2014) and Evans' groundbreaking research with UK householders suggests that 'worrying' or 'feeling bad' (2012, p. 46) are key sentiments attached to the disposal of food. Desires to escape the affective force of guilt and anxiety often set in motion a series of practices that while aiming to prevent waste, tend to create demands for more human household labour while simply delaying the surplus entering the waste stream. In the chapters that follow, I will examine the complexity of relations with food and food waste in households and among community organisations working to repurpose excess food. In these encounters, I identify how exposure to the affective force of this food waste (not simply at the point of discard but throughout its shifting states) and our necessary 'togetherness' with this waste may provide a basis for recognition of mutual vulnerabilities. These are risky encounters that often prompt forms of responsive, tinkering play. In seeing, touching and engaging with this material our reliance on nonhuman components to manage it are exposed. In so doing, our reliance, not only on technology or systems of service provision, but on relations among numerous entities is brought to the fore. It is in these spaces marked by the mutual vulnerabilities of becoming-with that Hird (2013) suggests new forms of ethics promoting attuned, responsive human/more-than-human interactions may be able to take root. Through the fieldwork explored in the following three chapters, I will attempt to highlight that such an ethics can be enabled through forms of playful tinkering and conceptualised as the ethico-political notion of convivial dignity.

In this scoping chapter, where I draw on wider literature on waste and reuse to set the scene for the ensuing focus on food, we shall see that such an 'ethics' is not necessarily driven by environmental sensibilities. Indeed, the invocation of 'green subjectivities' has been identified as an ineffective means of encouraging less resource-intensive lifestyles (Gregson, Metcalfe & Crewe, 2007; Head, 2016; Rowson, 2013; Waitt et al., 2012). Low-impact households themselves often don't indicate environmentalism as a key factor motivating their everyday practices (Evans, 2011; Waitt et al., 2012). In response to this, the cultivation of other life practices or 'cultural resources' (Head, 2016), such as frugality prompted by encounters with scarcity and abundance, are explored here as sites where recognition of mutual vulnerabilities can enable identification of alternative ways of being and doing in the world attuned to uncertain futures. To do this, I first discuss food flows through households, problematising the term 'waste' before briefly sketching out the scope of the food waste problem at the global level and how this is typically represented in policy and academic literature. Here, we will see the dominance of political-economic framings in response to which I argue for the necessity of attending to the ways in which waste is socially produced (drawing on the stretched notion of the social developed in Chapter 2) through visceral encounters. This is shown to enable identification of practices and narratives that can be drawn on to support the development of the skills and capacities needed to live in ways attuned to change and uncertainty.

Food flows: placing, removing and obscuring

Organic waste, much of it food, is estimated to be the largest contributor to rubbish in many nations around the world. Buried, burnt or otherwise removed from view, this excess does not simply disappear. Landfill is perhaps the most prolific marker of these efforts to obscure the ongoing becomings of waste (see Hird, 2009, 2013). The conduits that lead to these deep scars in the earth include the municipal services that facilitate their functioning and the minority world homes that feed them. As has been well-established (Chappells & Shove, 1999; Hetherington, 2004), household bins provide modes of habituating inhabitants into the rhythm of hiding and removal of waste in concert with the needs of the service providers. Bins that require the expending of more human labour to sort and separate waste within households (such as recycling and organic waste) can, according to Chappell and Shove, encourage people to 'identify with their wastes, to recognize their different constitutions, and to value some more than others' (Chappells & Shove, 1999, p. 278). However, while the narratives accompanying such services aim to encourage households to take responsibility for the local, national and global environmental impact of their waste (Chappells & Shove, 1999), the process of removal can also obscure its ongoing becomings. Indeed, as Gregson and Crang note, waste 'is a long way from stuff that "just is"... rather it becomes' (2010, p. 1028).

Neither the transition of goods to waste and eventual disposal from homes are inevitable processes, nor are they acts of finality. Waste does not 'end' or become stabilised as 'waste' when entombed in the earth. It continues to become. Moreover, while understandings of waste management practices often draw heavily on Douglas's work (2002) on dirt and pollution and systems of household or community ordering, Hetherington argues that such analyses fail to understand that 'the absent is only ever moved along and is never fully gotten rid of' (2004, p. 162). Staying with the small household scale, rather than attending to the bin as 'the archetypal conduit of disposal', Hetherington instead hails the door into this category due to its capacity to facilitate flow between commonly accepted binary categories (he writes of outside/inside, present/absent). Through this formulation, bins are not end points but rather the flows of their movement connect them to an ongoing journey—a moving along—of what is to become waste, to be buried in the earth or recovered and repurposed through laborious and often expensive recycling processes. These are ongoing processes that constantly push the perception of order and human control beyond reach.

The conduits and flows of wastes' ongoing journeys and the presence of absences highlight the inadequacies of the suggested finality commonly associated with the term waste. As such, within academic literature, we increasingly see a preference for notions of discard, as evident in the rise of discard studies. Hetherington exemplifies this conceptualising trend when he writes that waste 'suggests too final a singular act of closure' (2004, p. 159). Instead, he prefers to talk of disposal and acts of placing, terms that indicate the ongoing visceral impact of these goods that require bodies to act in response to the becomings of the materialities we encounter. In fact, the very vitality of waste (and what will become waste) produces a series of ongoing encounters among humans and nonhumans. For food waste, this is perhaps most powerfully and stealthily evident in the contribution to climate change of the methane produced by the rotting flesh of our what-was-once-food buried in landfill. But this is not food waste's only form of becoming in response to surplus and our management of it. The loss of water and phosphorous in the what-was-once-food are felt by future crops and the nonhumans and humans reliant on them for their lives and livelihoods. The rumbles in the tummies of the perpetually hungry highlight the presence of the absence induced by discarding food. However, as 'there is nothing inevitable' (Evans, 2012, p. 1125) about the transition, or modes of becoming, from surplus to waste, there are multiple points along the value chain where excess can become something else. The generative capacity of these sites of becomings harbours the potential to support more productive forms of human/nonhuman engagement.

The becomings of food waste

Food that is leftover or surplus rarely becomes waste immediately. Evans suggests that it often goes through 'a two-step' process before its value as food is lost and people reconcile themselves with the 'need' to discard it (Evans, 2012, p. 1130). This is often a means of alleviating, or trading off, guilt or anxiety. Once the what-was-once-food reaches a certain age or stage of decomposition that renders it undesirable, or worse, a health risk, its discard becomes a means of caring for self and others. In these ways, people's efforts to avoid waste through practices such as 'rescuing' uneaten leftovers by placing them in the fridge for later, while creating an initial 'diversion' from the waste stream, may not be successful in the longer term. In fact, food can get lost in the very bowels of the technology designed to prolong its life. Evans refers to some of these such as fridges and the Tupperware containers we populate them with as 'coffins of decay' (2012, p. 1132). While fridges are places that attempt to stabilise materialities, within them the very vitalities of food—or the multiple messmates with whom we live (Haraway, 2008b)—can make themselves seen, smelt and felt (Waitt & Phillips, 2016). These produce visceral, lively human/more-than-human encounters that encourage embodied responses to the excess of vitalities that we hoped the technology would contain: the smell of off milk, the fuzz of the green-grey mould colonising the plate of leftovers or the sloppy, stinky liquid in the plastic bag that once contained lettuce mix. The food now presents as a threat to the health of the family and the only viable option is disposal. Despite this trade-off, many remain worried about these acts of discard, haunted by the absence or the loss of the food's potential.

The graduated process through which food becomes waste is evidence that householders are not simply profligate consumers with little regard for the consequences of their excess. Instead, as Evans's notes 'households are following very specific procedures in order to manage the residual value of discarded foodstuffs and ameliorate anxieties about its wastage' (2012, p. 46). This is echoed in Waitt and Phillips' (2016) study of household fridge usage and ridding practices (part of a larger study exploring household sustainability) where people were found to invest a great deal of time in managing food circulation through their fridge via three key activities: placing, rotating and assessing. These hands-on practices can produce intimate, embodied engagement with food transitions guided by sensorial experiences as well as institutional and scientific governance (most overtly expressed in use-by dates). As Waitt and Phillips write '[r]epeated moving, touching and sighting of refrigerated foods allowed embodied understandings of foods' materialities and value' (2016, p. 370). These understandings, they suggest, can 'prompt productive moments in which participants may rethink their practices of food waste more broadly' (Waitt & Phillips, 2016, p. 364).

The generative aspects of these visceral engagements can include designing meals around decaying food (including rethinking the mode of consumption such as cooking vegetables instead of eating raw and stewing fruit that is deemed to be past its best), notes to avoid purchasing such large quantities in the future or a repurposing into animal food and compost. In these ways, surplus food presents as a site where the sensorially experienced entanglements of humans and more-than-humans induces affective responses that can prompt engagement in new consumption and discard practices. Through the small body of ethnographic writings we have on household food waste, we can see that food flows into, around, and out of, houses in complex ways often based on procedures enacted by householders that aim to nourish and support their health and wellbeing and manage concerns related to waste generation. These encounters can induce generative responses and thus deserve greater attention in our efforts to meet current demands for food waste reduction.

Encountering mutual vulnerabilities: the possibilities of convivial dignity

Even when the what-was-once-food passes through the various conduits to become discarded into mainstream bins leading directly to landfill, attention to the ongoing becomings of food can signal 'generative moments' of rethinking human/more-than-human entanglements centred on exposure to mutual vulnerabilities. Rotting food in landfill is no longer able to contribute to what we might call 'productive' nourishing relations, those that can support ongoing planetary life as we know it, feeding humans and nonhumans, from animals to soil microbes. However, the buried what-was-once-food does continue to provide sustenance, now metabolised by bacteria that produce uncontainable leachate likely to do harm to many in our relational entanglements (Hird, 2009, 2010, 2013). It is the largely unsensed reality of these microbial relations that, for Hird, exposes vulnerabilities that could provide the basis for a new form of 'environmental ethics' (2013, p. 107). As Hird writes, 'bacteria remind us that most relational encounters on earth have nothing to do with humans; nor are humans even aware of most of these encounters and assemblages' (Hird, 2013, p. 110).

Bacterial relations challenge visions of control and superiority associated with anthropocentric minds, bodies and narratives. For Hird (2013), such exposure lays the groundwork for an ethics of vulnerability that is not predicated on intimacy, shared recognition or empathy but on the necessity of our being together in sometimes unsettling mutual entanglements with entities often unknown and unknowable to each other. Such a position is predicated on an ethics of responsibility, one embedded in noticing and attuning; one that, at its core, is about humans encountering relations that are not only beyond their control, but micro-ontologies where humans do not figure at all. Throughout this text, I have suggested that this sensing and recognition of our necessary togetherness-in-relation induced by mutual vulnerabilities can be understood through the ethico-political notion of convivial dignity. I develop this as an intentional 'linguistic jolt' that takes up the challenge of contributing to the development of alternative grammars that aim to disrupt anthropocentric narratives, and the modes of being and doing these support, by stretching conceptions of dignity not only to more-than-humans but beyond bounded entities (beyond even the in-flux state of many microbes) to the very relations through which life and matter congeal and shift.

Attuning to the becomings of our surplus food, and initiating and taking up invitations to play with these, provides rich opportunities to encounter these relations in new ways. Indeed, while landfill, like digestion, attempts to obscure human/nonhuman relational becomings, the visceral encounters with its impacts—most palpably through leachate's contamination of soil and water supplies, methane's contribution to climate change and encounters with peak capacity of landfill—means these efforts inevitably fail. Thus, the presence of this absence continues to ensure food waste has the capacity to prompt more relational, attuned approaches to togetherness-in-relation. However, this is not how waste is typically addressed.

Conceptualising food waste

In the belly of the modern capitalist minority world, dominant approaches to waste centre on issues of governance and management, with food waste typically viewed through a political economic lens (Busch & Bain, 2004; Campbell, 2009;

Freidberg, 2004a, 2004b; Stuart, 2009). In regard to governance, food waste has received growing international attention in recent years. Reduction efforts have particularly focused on commercial and household levels with emphasis placed on the economic, environmental and social justice issues revolving around excess food. This is exemplified in the Sustainable Development Goals (SDG) where a subset of the 12th goal to 'Ensure sustainable consumption and production patterns' aims, by 2030, to 'halve per capita global food waste at the retail and consumer levels and reduce food losses along production and supply chains, including post-harvest losses' (UN, 2016, p. 27). The US Government has introduced the same target and the French Government has banned supermarket disposal of still edible foods. Outside of these international goals and national regulatory systems, growing numbers of commercial enterprises such as supermarkets have also initiated plans to reduce food waste and to be held to account for their waste generation by providing publicly accessible waste statistics. For example, the UK chain Tesco has committed to the latter, indicating a four per cent rise in food waste in 2015 (Wood, 2016). Woolworths, one of two main Australian supermarket chains, has mapped out a plan to send zero waste to landfill by 2020, though it failed to achieve its 2015 waste reduction target (Han, 2015).

These ambitious commercial food waste reduction goals and growing commitment to transparency in reporting food flows have been hampered by a lack of consistently defined, collected and verified global food waste data. Indeed, estimates range rather widely from between ten and forty per cent of the world's total food production being lost at some stage of the food system, from production to consumption (Parfitt, Barthel, & Macnaughton, 2010, p. 306). In part, the accuracy of the data is fuelled by divergent approaches to understanding and defining what food loss and food waste actually are. In response to this lack of reliable data and growing recognition of the economic, social and environmental costs of food waste, in 2016 the first global standards to measure food loss and waste, the Food Loss and Waste Accounting and Reporting Standard—or FLW standard (Hanson et al., 2016a)—and the establishment of universal definitions to assist this process were developed by a group of international organisations from the EU, UK and the UN brought together by the World Resources Institute (known as the Food Loss and Waste Protocol). While Gille notes that 'leaving data unreliable or raising doubt about calculations is a key tool in delaying regulation' (Gille, 2012, p. 38), here we see somewhat of the reverse. As nations such as France and the US take national regulatory action, the need for reliable data has increased.

However, efforts to reach agreement on definitions have failed to radically query the positioning of waste as an end-point, an entity marked by its very finality. The lack of attention to the ongoing process of the becoming of waste may well be a factor that hampers international efforts to reduce food waste. The limits of this approach are hinted at in the FLW's standard's definition of inedible food where it notes that 'what is considered inedible varies among users…, changes over time, and is influenced by a range of variables including culture, socio-economic factors, availability, price, technological advances, international trade, and geography' (Hanson et al., 2016b, p. 2). Indeed, waste's becomings, following Haraway, would be better framed as becomings-with and are produced by multiple factors including cultural, economic, social and spatial issues. As Gille points out, the logic fuelling the differentiation between food loss and its implied potential to be saved and food waste which indicates a sense of finality serves to encourage a localised misconception of the social, economic, political and cultural entanglements that lead to food becoming unavailable to eat (Gille, 2012).

While around the same percentage of food is estimated to enter the waste stream in both the minority and majority worlds, in the latter most food is identified as being lost or spoilt along the supply chain from pest-inflicted damage at the site of production due to poor storage and transport problems (Gille, 2012; Gustavsson, Cederberg, & Sonesson, 2011) and is commonly known as food loss. Such issues have mostly been eliminated in the developed world with food waste occurring in these more affluent economies primarily through the distribution and point of sale components of the food system and generally referred to as food waste. Gille asserts that analysis, understanding and potential reduction of food waste can only occur if we pay adequate attention to the social relations and multi-scalar networks that inform the very production or becomings-with of waste (Gille, 2012, p. 36).

The socio-technical creation of waste

Indeed, conceptions of economic, technological, legal, political and biological risk across multiple scales, from local to global, are key drivers of waste creation (Gille, 2012). Gille (2010, 2012) writes of 'food waste regimes' (which apply to the production, representation and politics of waste) whereby efforts to minimise risk effectively produce waste at particular points in the conduits through which food travels, reducing what is available, and able, to be consumed (Gille, 2012, p. 37). Indeed, she demonstrates that 'current thinking on food waste does not merely reflect risk avoidance strategies, but also how that thinking affects how we actually produce food waste' (2012, p. 37). This is echoed by Krzywoszynska when she identifies that the movement of materials into the category of waste 'is not necessarily linked with the politics of value, or indeed with environmental concerns' but rather that legislation attending to waste is a 'uniquely powerful regulatory intervention into the spaces of agro-food production as sites of risk, be it environmental or economic' (Krzywoszynska, 2012, p. 48). This can include waste produced in a multiplicity of ways, such as via rejection of fruit and vegetables at the farm gate that fail to meet the aesthetic standards of supermarkets, by-products of farm production, consumer misunderstanding around best-before dates coupled with overly cautious approaches to use-by dates, and via efforts to ensure adequate world food supply and minority world farmer income through overproduction.

The latter is exemplified by the case of US food aid flooding markets in the majority world, consequently reducing demand for locally grown produce (and sometimes threatening export to other markets as we saw in relation to GM foods in Zimbabwe—see Gille, 2012; Herrick, 2008; Turner, 2011; Zerbe, 2004). This example highlights power inequalities but also that 'we live in a food waste regime in which there is not only a mechanism for bad weather and pests to result in food waste, but also a mechanism in which good natural conditions lead to waste as well' (Gille, 2012, p. 34). This regime and its outcomes emphasise the significance of the social context within which food becomes waste. As Gille argues, 'waste itself—its production, its consumption, its circulation, and metamorphosis—is constitutive of society' (2010, p. 1050). Attention to these processes, and the affective force of waste, are important sites of focus for food waste reduction endeavours.

The affective force of visceral encounters with food waste

Waste troubles us. Acuto asserts that it is its 'assembled nature', comprising 'nonhuman and human components' across multiple scales that 'makes waste so wicked' (2014, p. 351). Intense embodied responses to waste, such as disgust and revulsion, also function as 'powerful political forces' that demonstrate the 'political possibility of affect as a response where different becomings might emerge' (Hawkins & Muecke, 2003, p. xiv). To mobilise these affective forces, perhaps we need something akin to Hawkins' new ethics of waste that works against existing food waste regimes. Such an ethics would facilitate and celebrate more creative, engaged and closer relationships with our waste (Hawkins, 2006). We see something of this ethics in dumpster divers, many of whom are adherents of freeganism (an anti-consumerist ideology which encourages people to limit their participation in neoliberal economic relationships). For freegans who dumpster dive, the act of repurposing discarded food can be understood as a political and ethical act aimed at intervening in the inefficiencies of the global agri-food industry (Edwards & Mercer, 2013; Turner, Henryks & Pearson, 2011). Freegans are intimately engaged with reinterpreting what the food system discards as waste. Indeed, they challenge the logic of use-by and best-before dates imposed by the governance of the industrial agri-food system instead using 'their innate senses of touch, taste and smell' to inform their food choices, thus taking responsibility for their own food safety (Edwards & Mercer, 2007, p. 290). This form of embodied engagement with what some parts of the food system identify as waste may provide a basis for a deeper engagement with food and broader understanding of our human/more-than-human entanglements.

The use of senses to determine the freshness or edibility of food was all humans had to rely on prior to the industrialisation of the food system. It remains so for many in the developing world today and some marginalised groups in the developed world. The use of senses to assess the suitability of food is key to animal survival. For humans, such skills seemed to have waned in urban settings in the developed world where many have come to rely on the risk assessment practices built in to the industrial food system to make our food safe. Even for those who do tend to draw on their senses to determine food safety within their homes (most commonly by smelling milk on or past its use-by date), this assessment works alongside the scientific and regulatory information provided rather than as a direct replacement (see Waitt & Phillips, 2016). A rethinking of waste, a reengagement with these senses and the development of more embodied, visceral relationships with food and its becomings may be practices capable of generating new forms of understanding of human/more-than-human entanglements.

As we have seen in earlier parts of this text, such forms of visceral attunement are, when coupled with particular affordances, capable of supporting engagement in interactions where mutual vulnerabilities are identified and tinkering modes of play engaged in. Through these practices, recognition of the inescapability of our relational entanglements or togetherness-in-relation come to the fore. Responsiveness to this togetherness means that those involved aren't only moved by matter but by the propositional and generative potential of the relations that enliven it. This relational responsiveness resonates with generative potential for the development of alternative ontologies that can support less resource intensive ways of being and doing with the world. However, these practices are challenging to narrativise due not only to the recalcitrance of matter and relational flows but as a result of the dominance of anthropocentric narratives. As such, I contend that the challenge to promote alternative forms of the modern human subject that lives a more sustainable life not only require attention to matter but also to discourse. These are material-semiotic challenges. Convivial dignity is one attempt to provide a grammar for representing the possibility that life and relations among human and more-than-humans exceed the sum of their parts. These are not relations that rely on enactment of care or the forging of attachments. Instead, multisensorial encounters with waste highlight the messiness of our relational entanglements. This often manifests in responsive practices of making do that can be heightened in times of scarcity and abundance.

The relational becomings of making do

Of course, my focus on the generative potential of waste relations is not equivalent to calls to encourage people to engage in dumpster diving. The growing raft of legislation preventing the binning of still saleable food may move this practice onto other more 'acceptable' sites of food redistribution, likely to be in the form of charities required to then rehome the food. While the vulnerabilities of the dumpster divers ('will this make me sick'?) are palpable in the fresh foods' rapid becomings, aided and abetted by a host of microorganisms, engagement with surplus food also invokes a range of other vulnerabilities for humans and nonhumans. For the former, food waste—and waste in general—are often identified as forms of unnecessary monetary loss that householders identify as occurring when their efforts to manage their household efficiently fail.

As indicated above, waste is something that often prompts feelings of unease and guilt. This is often induced by householders' attempts to practice thriftiness or frugality whereby unnecessary consumption is avoided and people devote time and effort to repurposing goods to avoid consigning them to landfill. This is commonly linked to experiences of a childhood where people recall not having much (often something they only become aware of on reflection, rather than being cognisant of at the time of their actions) and where the general approach was one of making do. As such, overconsumption is linked to vulnerability to poverty but also to scarcity in a more general sense. Those who tend to 'make do' often identify concerns, not simply about potential future limitations to their purchasing capacity but about the finite limits of resources and the inputs that produce them.

Here, it is important to clarify the ways in which I am mobilising the notions of thrift and frugality. Watson and Meah (2013, p. 113) define thrift as being 'concerned with responsible and conservative use of resources'; however, Evans links thrift with the hunt for 'bargains' so that more economic resources can be freed up for ongoing purchasing to enable care (highlighting a moral dimension of this activity) for self and loved ones (Evans, 2011, p. 551). Thriftiness in this rendering is entangled with consumption and little attention is paid to vulnerabilities to broader resource scarcity. The scope of concern is quite narrow with the emphasis on care for immediate loved ones. While attention to the wellbeing of those close to us can also be a feature of frugality, Evans defines it more broadly as a social practice that involves being 'moderate or sparing in the use of money, goods and resources with a particular emphasis on careful consumption and the avoidance of waste' (Evans, 2011, p. 552).

If we take this to be the case, then frugality is not simply driven by the dollar. Frugal practices can include avoiding cheap items if more expensive variations are considered to be longer lasting or capable of providing a wider variety of uses. Frugality is driven by desires to avoid waste and its practitioners attempt to ensure that goods (from food to TVs) and services (water to electricity) are made best use of. It is excess, unnecessary usage and acquisition of too many things that is avoided. As such frugality is not necessarily a response to economic scarcity. The distinction between thrift and frugality is significant to our efforts to determine which 'cultural' resources can be drawn on to promote less resource intensive lifestyles, and I adopt Evans' definitions throughout this text.

Mobilising ethico-political change: the limits of an environmental ethic and economic vulnerabilities

While freegans are staunchly anti-capitalist and environmentally minded, practitioners of frugality and thriftiness do not necessarily adopt a particular politico-ethical stance even when their behaviours and habits can be readily identified as being sustainable. There is a common sentiment, anecdotally and in much of the literature, that invoking environmental sensibilities or 'green subjectivities' is not the most useful way of encouraging broad uptake of changes (Gregson et al., 2007; Head, 2016; Rowson, 2013; Waitt et al., 2012) nor is it the ethic that is commonly expressed as motivating less resource-intensive lifestyle by the actual practitioners themselves (Evans, 2011; Waitt et al., 2012). Indeed, Head contends that the very characteristics needed to live in the Anthropocene do not require a 'green' mentality (2016, p. 13). She suggests that practices of thrift, and thus a focus on economic benefits for households, may be a useful place to focus attention to encourage more conservative use of resources. However, for Evans

(2011), the link between thrift and environmentally sustainable habits is much more fraught. By drawing on historical analyses, he details that while some expected the recent economic downturn to prompt more sustainable consumption it was likely to encourage bargain hunting rather than reduced consumption, noting that 'the passage from care over scarce economic resources to care for the environment may not be as smooth or clear as might be hoped for' (Evans, 2011, p. 556). This is a sentiment informed by Trentmann's claim that 'scarcity does not automatically trigger a caring instinct [...] any more than affluence automatically leads to indifference' (2011, p. 556).

Indeed, economic vulnerability on its own is unlikely to be sufficient to prompt broad scale behavioural changes, but concerns about, and direct experiences of, scarcity (and, as we will see below, also abundance), may be more generative of new forms of resource-use behaviours and waste ethics. As indicated above, frugal practices are not necessarily driven by financial vulnerability. They are often also attuned to concerns about broader resource availability and the desire to avoid waste. These, however, do not have to be couched in relation to notions of care, which, as argued in earlier chapters, may not provide the most generative prompts for ecologically sensitive practices. Care in times of scarcity may not be attuned to an ecological sensibility but care for self and family that could induce environmentally damaging behaviours. Attention to just how and why waste is avoided outside of a focus only on economic or environmental reasons is needed to assist us to identify key ways of reducing resource usage and altering broader anthropocentric behaviours. Some of these appear to develop in a playful manner marked by tinkering.

The affective force of lively surplus: playful practices of waste diversion

Many practices aimed at reducing waste are already common in people's everyday lives. By understanding the very capacities and affordances that enable these to occur we may be able to identify ways of promoting and supporting these behaviours. People commonly experiment with modes of redistribution and reuse, for example regifting among family, selling on ebay, being part of garage-sale trails and donating to charity (Bulkeley & Gregson, 2009; Gregson, Crang, Laws, Fleetwood & Holmes, 2013; Hitchings, Collins & Day, 2015). While the shifting vitalities of food introduce temporal and safety issues to the modes of repurposing, increasingly we are seeing the development of modes of repurposing that extend beyond the household that generated it, for example food sharing apps such as Olio. Participation in these often time-consuming practices indicates a reluctance to waste and a desire to extend and expand the life of material goods. It also speaks to a playful, tinkering approach whereby the householders are responsive to the materials themselves and the shifting affordances—namely the use of online forums and recent neighbourhood initiatives such as the garage sale trail and sharing/buy/swap/sell apps—that enable them to be moved on.

This can, for some, be partially economically driven, but existing research (not yet including food examples) indicates that those who invest in these practices display a strong drive to avoid committing usable objects to landfill. Indeed, Hitchings et al. found that this was 'largely an outcome of senses of gratitude and responsibility' (2015, p. 377) produced by a social and familial context which encouraged recognition of the value of things. This may well be the flip-side of the guilt and shame that can accompany new consumption or waste (Evans, 2012). Surplus goods, along with that which eventually becomes waste, are affective in force. This affect moves things along in various ways and these contexts contribute to the development of subjectivities whereby people are marked by their engagement in practices that distance them from being members of a profligate, throwaway society.

Through the myriad of householders' practices of reuse and moving along conduits to avoid goods transitioning into waste, we can discern a sense of object vitality. The material goods, while surplus to needs, are recognised as still having 'life' left in them for others or other purposes. Such practices may not constrain ongoing consumption (indeed, the passing on of goods is usually done so as to enable the purchase of a newer replacement model), but careful attention to these modes of repurposing do indicate stirrings of attunement to materiality. This is more closely aligned to the notion of frugality than thrift as outlined by Evans (2011). Indeed, practices of waste avoidance in frugality are marked by attunement and responsiveness to the very materiality of goods and their vitality and potential. Engagement in these practices can encourage the training of sensitivities to better enable recognition of vitality. Whether motivated by gratitude and responsibility, attempts to stave off guilt and shame, or a combination of the two, reuse and repurposing highlight human vulnerability to the affective force of material goods and can encourage less resource-intensive lifestyles. As indicated earlier, this capacity to be moved by more-than-humans does not need to be motivated by a 'green ethic' or financial incentives (Head, 2016). Other ethical values related to social contexts may be at play here.

Social contexts, rather than individual moral persuasions, have been identified as key drivers of resource conserving lifestyles (Bulkeley & Gregson, 2009; Hitchings et al., 2015). This position challenges typical waste policy approaches that are 'firmly tied to knowledge (awareness raising) and the hip-pocket nerve (financial incentives) as the primary barriers to/drivers of action' (Bulkeley & Gregson, 2009, p. 934), with Bulkeley and Gregson instead emphasising the importance of social, physical, spatial and material aspects (2009, p. 935). As noted above, households engage in myriad practices to reduce waste, particularly through passing things on for reuse (including donations to charity and sharing among familial and social circles) to such an extent that Bulkeley and Gregson note '[r]euse, then, be it marked by monetary exchange or not, is an integral part of how most UK households think about the surplus of consumption' (2009, p. 938, emphasis added). However, the capacity to engage in these behaviours is informed by social networks and spatial aspects related to accessibility and social acceptance. As such, efforts to reduce waste need to attend to these factors by acknowledging the complexity of the conduits goods pass through before becoming, or being diverted from becoming, waste. There is a need to move beyond reliance on assumptions that we inhabit a 'throwaway society' and that overconsumption should be addressed solely through economic incentives and mobilisation of environmental sensibilities.

Retooling resource use: scarcity, abundance and material limits

Hitchings et al. (2015) speak of 'inadvertent environmentalists' in their research into how older people keep warm in winter and how younger people attend to surplus goods. These sustainable resource users tend to be fuelled by 'values' related to responsibility and care for others and what are seen to be common sense approaches. However, Hitchings et al. (2015) also suggest that highlighting the greenness of these practices may enable people to begin to identify themselves as 'sustainability champions' which could encourage extension of ecological sensibilities to other areas of their lives (2015, p. 380). While the utility of such approaches is context specific, attention to the characteristics of the affordances that support less resource-intensive lifestyles is important for encouraging and supporting greater uptake of these behaviours. For example, for older people, memories of disruption to resource supplies during wars and rationing are commonly identified as key to supporting the development and enactment of the skills and capacities to live less resource-intensive lifestyles attuned to uncertainty.

Experiences in contexts marked by change and disruption can support human adaptability to less resource-intensive lifestyles marked by reduced waste. This is particularly apparent in times of scarcity and abundance. The visceral experiences of scarcity—of a lack of resources that push different ways of living beyond reach through encounters with material limits—for example, in times of drought power outages and food rationing forces adaptations that can draw on and extend the skills and capacities of those involved (Head, 2016; Strengers & Maller, 2012). Strengers and Mallor's work on water and energy practices in Australian migrant households found that an emphasis on 'resource characteristics of materiality, diversity and scarcity' (2012, p. 754) enables identification of householder capacities to adapt to changes in access to resources. In particular, they identify scarcity as a prompt for transforming household practices. Once again, these are not necessarily behaviours that are induced by economic necessity, a green ethic or 'moral' imperatives of care and compassion (Evans, 2011). Instead, these changes are brought about by experiences of shared vulnerabilities, of having to 'live with' and 'make do' in the face of change and disruption wrought by scarcity. Here, people are regularly confronted with having to let go of things they have previously cared about or been attached to. These are encounters that often demand the development of new relations often with unknowable others. This can lead to engagement in experimental practices and what I conceive of as forms of responsive playful tinkering that often involves recognition of the necessity of togetherness-in-relation and, thus, could be marked by forms of convivial dignity.

Abundance can also bring these mutual vulnerabilities to the fore via encounters with material limits and agency. This can be seen, for example, through flooding (too much water) or the production of surplus crops in a saturated market.

Exposure to these vulnerabilities acts as a cue for the development and adaptation of practices attuned and responsive to material conditions and the relations that bring these into being. This also often incites playful tinkering. It can enhance, or turn people into 'makers' envisaged by Carr and Gibson as those who carry out 'a series of negotiations and concessions with the material, working within a realm of possibilities that are afforded by its particular properties' (Carr & Gibson, 2016, p. 303). In this way '[m]aking becomes a material conversation—a physical provocation and a response, iterated over and over again, working with the material to understand its capacities, analyse error and make adjustments' (2016, p. 303). Through fine-grained analysis of the making practices of steel workers, Carr and Gibson demonstrate that industrial-borne skills and haptic engagement with materials which enables creative, resourceful repairs, remakings and reworkings may provide 'ironically, an ability to cope with volatile futures' (2016, p. 307).

Here, we can see capacities and affordances attuned to vitality, as Carr and Gibson write, these are '[p]eople who are skilled in dealing with the material world in the face of disruption' (2016, p. 307). Thus, unlike economic imperatives (where value is assigned by capitalist infrastructure and financial incentives to conserve resources, often to enable a diversion of funds to other purchases) and traditional conservationalist mentalities of care and protection (which I argued in earlier chapters are often representative of anthropocentric attitudes), being responsive to mutual vulnerabilities by making do and acknowledging the necessity of togetherness-in-relation brings the more-than-human and relational flows into view and encourages playful engagement responsive to mutual entanglements through the enactment of convivial dignity.

Conclusion

Multisensorial, attuned engagements with surplus highlight the becomings-with of waste. While those of us in the minority world live in societies that by and large do their best to obscure the ongoing transformations of organic waste and the mutual vulnerabilities these processes draw attention to, visceral, intimate encounters with this matter-in-relation can, with the right affordances, have considerable affective force promoting recognition of human/more-than-human relational entanglements. This can support an openness to being moved by matter and the propositional and generative potential of the relations that enliven it that can manifest in myriad practices of making do. This recognition of togetherness-in-relation is capable of forming the basis for alternative ways of being and doing in the world that resist anthropocentric norms and contribute to enactment of less resource intensive lifestyles.

Engagement with, and extension of the life of, material goods through forms of playful tinkering do not have to be predicated on forms of environmentalism. They can be prompted by sensibilities related to other moral imperatives (such as frugality, see Evans, 2011; Head, 2012), concerns about 'potential social stress' (not having clothes or devices, see Hitchings et al., 2015) as well as temporal and financial issues. It is the process of encountering and engaging anew—entering into relations with unknowable others and being open to being moved by the resulting relations through enactment of convivial dignity—that we can identify the generative potential for alternative ontologies and world-making practices. Food waste, surplus and discard provide sites ripe for material-semiotic reconfigurations that may be capable of challenging the dominant, environmentally destructive ways these becomings are attended to in the minority world.

In the following three chapters, I test out these tentative potentialities through analysis of fieldwork that attends to the management of food in peoples' homes, gardens and backyards, and through food rescue organisations and the ugly food movement. Here, we will see how playful interactions with leftover, surplus or wasted food can assist in 'training our sensitivities' (Mol, 2009, p. 278) to become more attuned to the vitality and affective forces of the material relations within which we act and are enacted. In these sites, we will encounter people who are developing the skills and capacities to adapt to future uncertainty and the enactment of alternative ontologies capable of supporting more sustainable futures. While taste seems to lend itself to these generative playful, convivial forms of engagement, extension of these concepts to waste is much more experimental.

However, if those who eat together stay together, I wonder what opportunities for convivial living between humans and more-than-humans we might be able to glimpse if we recognise and cultivate social aspects of waste—specifically in relation to food—instead of continuing our current solitary efforts to contain it through removal from our homes in plastic bin liners to be entombed within the earth in landfill. As we will see, convivial relations are not always harmonious. Recognition of vitality in the material world and the relations through which these are formed does not necessarily induce enchantment or awe. It can lead to frustration, disgust, dreams of control and desires for detachment. But its affective force has the potential to induce recognition, or sensing, of forms of togetherness-in-relation.

In the next chapter, these ideas will be enlivened through dwelling with the waste minimisation strategies and the flows food takes through the homes of gardeners and AFN participants. The food related practices these participants undertake are shown to be motivated by recognition of multispecies entanglements and realisation of mutual vulnerabilities that prompts the gardeners and shoppers to be open to being moved by food. This responsiveness to the becomings-with of food is shown to be supported by the affordance of uncertainty, particularly as experienced in relation to encounters with food abundance and scarcity. The resulting recognition of togetherness-in-relation encourages creative engagement with food developed in a playful, tinkering mode that is underpinned by growing openness to relational flows marked by convivial dignity. This is shown to manifest in limited production of food waste, even though these contingent practices often deviate significantly from those promoted in food waste reduction campaigns.

Chapter 9 takes us deeper into the world of the relational becomings-with of food waste through its attention to compost heaps, bokashi buckets, worm farms and backyard chooks in the homes of gardeners. Here, the transformational shifts of food waste are found to be vibrant and vital in their multispecies abundance. Compost is shown to be a site rife with forms of togetherness-in-relation that evade the forging of simple relations of care, attachment or detachment but which are encountered through, and prompt engagement in, playful tinkering. The openness to living with and being responsive to the uncertainty and unknowability of these sites of slippery relations can be understood as being guided by convivial dignity which is shown to provide a basis for the enactment of non-anthropocentric modes of being and doing.

The final chapter in this section broadens our sphere of concern from the intimacy of homes and backyards to include more public sites of engagement with surplus food through encounters with food redistribution organisations and discussion of the global ugly food movement. Here, I identify that responsiveness to the becomings of food may not support reconfigurations of the modern human subject and enactment of playful tinkering guided by convivial dignity when scaled-up to macro levels of corporations. Indeed, in those sites, responsiveness is shown to enable a deferral of responsibility for the generation of food surplus. However, given the ongoing failures of governments and many corporations to take considerable action to limit their environmental damage, in this chapter I reinforce the necessity of attending to small-scale, local actions in our efforts to counteract anthropocentric-fuelled destruction. Through discussion of workers and volunteers involved with food rescue, I once again demonstrate how practices of making do—seeing 'life' left in surplus food or recognising the potential for it to be reconfigured, speak to engagement in practices of playful tinkering that can support attunement to togetherness-in-relation guided by convivial dignity.

References

Acuto, M. (2014). Everyday international relations: Garbage, grand designs, and mundane matters. *International Political Sociology*, *8*(4), 345–362.

Baker, D., Fear, J., & Denniss, R. (2009). *What a waste: An analysis of household expenditure on food*. Canberra, ACT: The Australia Institute

Bulkeley, H., & Gregson, N. (2009). Crossing the threshold: Municipal waste policy and household waste generation. *Environment and Planning A*, *41*(4), 929–945.

Busch, L., & Bain, C. (2004). New! Improved? The transformation of the global agrifood system. *Rural Sociology*, *69*(3), 321–346.

Campbell, H. (2009). Breaking new ground in food regime theory: Corporate environmentalism, ecological feedbacks and the 'food from somewhere' regime? *Agriculture & Human Values*, *26*(4), 309–319.

Carr, C., & Gibson, C. (2016). Geographies of making: Rethinking materials and skills for volatile futures. *Progress in Human Geography*, *40*(3), 297–315.

Chappells, H., & Shove, E. (1999). The dustbin: A study of domestic waste, household practices and utility services. *International Planning Studies*, *4*(2), 267–280.

Douglas, M. (2002). *Purity and danger: An analysis of concepts of pollution and taboo* (2nd ed.). London and New York: Routledge.

Edwards, F., & Mercer, D. (2007). Gleaning from gluttony: An Australian youth subculture confronts the ethics of waste. *Australian Geographer*, *38*(3), 279–296.

Edwards, F., & Mercer, D. (2013). Food waste in Australia: The freegan response. *The Sociological Review*, *60*(S2), 174–191.

Evans, D. (2011). Thrifty, green or frugal: Reflections on sustainable consumption in a changing economic climate. *Geoforum*, *42*(5), 550–557.

Evans, D. (2012). Binning, gifting and recovery: The conduits of disposal in household food consumption. *Environment and Planning D: Society and Space*, *30*(6), 1123–1137.

Evans, D. (2014). *Food waste: Home consumption, material culture and everyday life*. London and New York: Bloomsbury.

Freidberg, S. (2004a). The ethical complex of corporate food power. *Environment and Planning D: Society and Space*, *22*(4), 513–531.

Freidberg, S. (2004b). *French beans and food scares: Culture and commerce in an anxious age*. Oxford: Oxford University Press.

Gibson-Graham, J. K. (2011). A feminist project of belonging for the anthropocene. *Gender, Place and Culture*, *18*(1), 1–21.

Gille, Z. (2010). Actor networks, modes of production, and waste regimes: Reassembling the macro-social. *Environment and Planning A*, *42*(5), 1049–1064.

Gille, Z. (2012). From risk to waste: Global food waste regimes. *The Sociological Review*, *60*(S2), 27–46.

Gregson, N., & Crang, M. (2010). Materiality and waste: Inorganic vitality in a networked world. *Environment and Planning A*, *42*, 1026–1032.

Gregson, N., Crang, M., Laws, J., Fleetwood, T., & Holmes, H. (2013). Moving up the waste hierarchy: Car boot sales, reuse exchange and the challenges of consumer culture to waste prevention. *Resources, Conservation and Recycling*, *77*, 97–107.

Gregson, N., Metcalfe, A., & Crewe, L. (2007). Identity, mobility, and the throwaway society. *Environment and Planning D: Society and Space*, *25*(4), 682–700.

Gustavsson, J., Cederberg, C., & Sonesson, U. (2011). *Global food losses and food waste*. Paper presented at the Save Food Congress, Düsseldorf, Germany. Retrieved from http://www.madr.ro/docs/ind-alimentara/risipa_alimentara/presentation_food_waste.pdf

Han, E. (2015, September 25). Woolworths misses food waste target but sets new goal with OzHarvest partnership. *The Sydney Morning Herald*. Retrieved https://www.smh.com.au/business/companies/woolworths-misses-food-waste-target-but-sets-new-goal-with-ozharvest-partnership-20150921-gjr7k7.html

Hanson, C., Lipinski, B., Robertson, K., Dias, D., Gavilan, I., Gréverath, P., . . . Quested, T. (2016a). *Food loss and waste accounting and reporting standard*. Washington, DC: World Resources Institute.

Hanson, C., Lipinski, B., Robertson, K., Dias, D., Gavilan, I., Gréverath, P., . . . Quested, T. (2016b). *Food loss and waste accounting and reporting standard: Executive summary*. Retrieved from http://www.wri.org/sites/default/files/FLW_Standard_Exec_Summary_final_2016.pdf

Haraway, D. (2008a). Otherwordly conversations, terran topics, local terms. In S. Alaimo & S. Hekman (Eds.), *Material feminisms* (pp. 157–187). Bloomington, IN: Indiana University Press.

Haraway, D. (2008b). *When species meet*. Minneapolis and London: University of Minnesota Press.

Hawkins, G. (2006). *The ethics of waste: How we relate to rubbish*. Lanham, Maryland: Rowman & Littlefield Publishers.

Hawkins, G., & Muecke, S. (Eds.). (2003). *Culture and waste: The creation and destruction of value*. Lanham, Maryland: Rowman & Littlefield Publishers Inc.

Head, L. (2012, June 26). Australia's rich talk about saving the environment; the poor bear the burden of doing it. *The Conversation*. Retrieved https://theconversation.com/australias-rich-talk-about-saving-the-environment-the-poor-bear-the-burden-of-doing-it-7693

Head, L. (2016). *Hope and grief in the anthropocene: Re-conceptualising Human–Nature relations*. Oxon and New York: Routledge.

Herrick, C. (2008). The Southern African famine and genetically modified food aid: The ramifications for the United States and European Union's trade war. *Review of Radical Political Economics*, *40*(1), 50–66.

Hetherington, K. (2004). Secondhandedness: Consumption, disposal, and absent presence. *Environment and Planning D: Society and Space*, *22*(1), 157–173.

Hird, M. J. (2009). *The origins of sociable life: Evolution after science studies*. Houndmills, Basingstoke: Palgrave Press.

Hird, M. J. (2010). Meeting with the microcosmos. *Environment and Planning D: Society and Space*, *28*(1), 36–39.

Hird, M. J. (2013). Waste, landfills, and an environmental ethics of vulnerability. *Ethics and the Environment*, *18*(1), 105–124.

Hitchings, R., Collins, R., & Day, R. (2015). Inadvertent environmentalism and the action–value opportunity: Reflections from studies at both ends of the generational spectrum. *The International Journal of Justice and Sustainability*, *20*(3), 369–385.

Krzywoszynska, A. (2012). 'Waste? You mean by-products!' From bio-waste management to agro-ecology in Italian winemaking and beyond. *The Sociological Review*, *60*(S2), 47–65.

Mol, A. (2009). Good taste: The embodied normativity of the consumer-citizen. *Journal of Cultural Economy*, *2*(3), 269–283.

Parfitt, J., Barthel, M., & Macnaughton, S. (2010). Food waste within food supply chains: Quantification and potential for change to 2050. *Philosophical Transactions of the Royal Society B*, *365*(1554), 3065–3081.

Rowson, J. (2013). *A new agenda on climate change: Facing up to stealth denial and winding down on fossil fuels*. London. Retrieved from

Strengers, Y., & Maller, C. (2012). Materialising energy and water resources in everyday practices: Insights for securing supply systems. *Global Environmental Change*, *22*(3), 754–763.

Stuart, T. (2009). *Waste: Uncovering the global food scandal*. New York: W. W. Norton & Company, Inc.

Turner, B. (2011). Embodied connections: Sustainability, food systems and community gardens. *Local Environment*, *16*(6), 509–522.

Turner, B., Henryks, J., & Pearson, D. (2011). Community gardens: Sustainability, health and inclusion in the city. *The International Journal of Justice and Sustainability*, *16*(6), 489–492.

UN. (2016). *Transforming our world: The 2030 agenda for sustainable development*. Paris, France: Author. Retrieved from https://sustainabledevelopment.un.org/content/documents/21252030AgendaforSustainableDevelopmentweb.pdf

Waitt, G., Caputi, P., Gibson, C., Farbotko, C., Head, L., Gill, N., & Staynes, E. (2012). Sustainable household capability: Which households are doing the work of environmental sustainability? *Australian Geographer*, *43*(1), 51–74.

Waitt, G., & Phillips, C. (2016). Food waste and domestic refrigeration: A visceral and material approach. *Social & Cultural Geography*, *17*(3), 359–379.

Watson, M., & Meah, A. (2013). Food, waste and safety: Negotiating conflicting social anxieties into the practices of provisioning. *The Sociological Review*, *60*(S2), 102–120.

Wood, Z. (2016, June 15). Tesco food waste rose to equivalent of 119m meals last year. *The Guardian*. Retrieved https://www.theguardian.com/business/2016/jun/15/tesco-food-waste-past-year-equivalent-119-million-meals

Zerbe, N. (2004). Feeding the famine? American food aid and the GMO debate in Southern Africa. *Food Policy*, *29*(6), 593–608.

8 Waste in the home

Introduction

In this chapter, we return to the homes of gardeners and AFN participants to explore how food waste is encountered and managed through the flows of food into, around and out of their homes. Here we find contact zones, where attunement to the multispecies becomings-with of food become viscerally apparent through multisensorial experiences with their transitions from what is conceived of as being edible to states of inedibility. Awareness of, and an openness to, the vitalities of food is shown to heighten these householders' recognition of mutual vulnerabilities which, while sometimes unsettling prompting desires for detachment, always induce recognition of the inescapability of togetherness-in-relation. These attunements are shown to be supported by two key visceral, embodied practices; firstly, via moving and being moved by food, and secondly, through responsiveness and adaptability to the contingencies experienced in times of both abundance and scarcity of food.

As identified in the taste section, I once again find that in the food waste minimisation practices of these householders, responsiveness to relational entanglements manifest in practices of making do that tend to unfold in a playful, tinkering manner. Here, matter is shown to be open to reconfigurations through shifting relations that prolong its 'life' in transformational and transformative ways. These actions and modes of being challenge notions of human exceptionalism and congeal in behaviours that minimise the production of food waste in these homes. I suggest that attentiveness to these alternative forms of reducing food waste have the potential to contribute to the skills and competencies necessary for adapting to contingent futures. However, as we shall see, these practices often run counter to the waste reducing action promoted in food waste reduction campaigns.

The affective force of food waste in homes

The huge amount of food waste generated in minority world households is regularly depicted in governmental and not-for-profit campaigns as evidence of the unthinking profligacy of a society marked by increasing affluence and access

DOI: 10.4324/9780429424502-8

to cheaper commodities. However, as outlined in the preceding chapter, depictions of a throwaway society inadequately represent over consumption and waste production and the extent to which waste troubles householders. People are engaged in complex relationships with food flows through their homes that are impacted on by social, economic, political, health and spatial issues. Rather than simply revisiting the effect of these broader imperatives and constraints, in this chapter I focus on exploring how 38 householders engaged in this research are impacted on by the very materiality and relational flows of the food that moves through their homes. The management of food, of course, cannot be removed from these bigger picture concerns. Indeed, Evans (2012, 2014) convincingly demonstrates that food choices in homes are not simply a result of individual preferences and choices. But, in this chapter, these engagements are explored in regards to the relational becomings of food and food waste in order to tease out the capacity for bodily practices attuned to materialities to lay the foundations for less resource-intensive lifestyles.

Evans' (2012, 2014) work demonstrates that while householders wish to avoid waste and experience guilt and feel bad when food waste occurs in their homes, there are multiple reasons why food becomes surplus to needs and enters the waste stream. These include infrastructural issues (such as not being able to buy sufficiently small quantities at supermarkets), wanting to provide adequate care for themselves and families by preparing 'proper' healthy foods (even when they suspect these will not be consumed) and the general 'flux' of life (travelling, having to go out to events etc). While these all seem like legitimate reasons for the creation of waste, such outcomes were not typical for the people who contributed to the fieldwork of this research. This may be because these Canberrans had already chosen to provision most of their food outside of mainstream food systems (through their own food production or via AFNs) and thus had already prioritised certain forms of food engagements in their lives. Yet, these people did also express similar challenges to managing household food as evident in Evans' (2012, 2014) research, namely time, family life, different food preferences among household members and the need to provide varied, healthy foods.

The Canberrans' level of attention and responsiveness to the materiality of food tended to counteract the force of other pressures in their weekly management of food meaning food waste was mostly avoided. This is not to imply that the householders in Evan's work could simply and easily 'choose' to act differently and reduce their food waste. The broader constraints gestured to above limit this capacity. Nor is my aim here to determine what forms of food engagements should be prioritised. Instead, I am interested in how low food waste producers enact their food management. The 38 Canberra householders in this work don't simply discuss desires for food waste avoidance, they often articulate an outright 'hatred' of waste. This sentiment can be discerned among Evans' (2012, 2014) cohort as well, but among the Canberrans this hatred seemed to have such affective force that it translated into a series of routines and bodily practices that significantly limited waste.

To tease out how this manifested, I now turn to these householders to explore how the relational materiality of the food itself informs and prompts their embodied routines. These participants come from a variety of socio-economic backgrounds with many identifying that they currently, or in the not-too-distant past, had to budget carefully to ensure adequate food was available in their households. However, at the time of the research, all households indicated that they were able to purchase and/or grow sufficient food. As such, the discussion of vulnerabilities throughout this chapter does not relate to risks of becoming food insecure.

Embodied hatred of waste: attuning to the generative potential of mutual vulnerabilities

The majority of people involved in this research exhibited powerful, emotive responses to the issue of waste. Most typically, this was expressed in the pronouncement 'I hate wasting food', which was quickly followed by details of how waste is avoided and what it feels like when waste occurs, as evident in Julie's response:

> I don't think I waste anything. I'm fairly ridiculously paranoid about wasting food [laughter]. I feel I'm letting myself down if I waste... I just think it's not a good thing! I don't think we should! Too much food's wasted, I think; and too much stuff's thrown away. I think it's really bad!

The feelings invoked by imagined, potential and actual waste were palpable and visceral and they prompted reflection on, and adaptation of, daily practices. Waste was seen as something that was not only avoidable through careful household management, reinforcing a neoliberal focus on householder choices, but something which 'good' people should avoid. The vehement hatred of waste, while sometimes fuelled by concerns related to economic loss and environmental impacts, was most commonly centred on the materiality of the objects themselves and their lost potential. Here, the ongoing becomings-with of waste, and human and societal roles in this process, were something people were highly attuned to, to the extent that when surplus did transition into the waste stream some experienced physical and psychological reactions. This was exemplified by Gabriella who noted:

> One of the other things that I hate is putting food in a rubbish bin. When I go to someone else's house it... physically distresses me. I even lived in a town house for a couple of years and we just had this great big kind of smelly rubbish bin on the deck with worms in it and holes in the top and stuff and I eventually carted it downstairs and off to someone's house. I just cannot put food waste in a rubbish bin.

Another interviewee, Judy, with strong memories of composting as a child, commented 'when you put food scraps in the bin it just feels weird because I know it's biodegradable and it could go somewhere else'. For these people, the act of sending food waste to landfill was not only resisted on a philosophical and ethical level but also in terms of bodily habits; their bodies resisted and felt strange when unable to facilitate the repurposing of food waste. In these examples, we catch glimpses of the ways in which the affective force of waste is embedded in bodily practices. Those identifying these positionalities exert considerable effort to avoid the bodily dissonance associated with discard into mainstream waste of materials that could become something else. The waste is identified as a resource capable of nurturing new 'productive' forms of life through other relational configurations, from worms to human food.

These responses to the affects of food waste indicate attunement to mutual vulnerabilities. The need for regular engagement with, and the rapid transitioning or multispecies and multimaterial becomings-with of, fresh foods intensifies experiences of these vulnerabilities perhaps more so than with other forms of waste that occur less frequently and where modes of disposal into mainstream waste are more difficult, such as hard rubbish. The affective force of the shifting relational materialities of foods implores these participants, physically and ethically, to avoid the bin. Where this is unable to occur, loss and frustration is experienced. While waste material will continue to become in ways that may well be harmful, such as through leachate in landfill, its potential to nourish non-destructive forms of life is lost.

However, the affective force of embodied engagements with food surplus and waste plays little role in the dominant campaigns that are used to promote food waste reduction. Before we further flesh out the Canberra householders' food relations, I want to take a brief look at the key food waste reduction behaviours advocated by one of the most well-known multinational campaigns, Love Food Hate Waste. In so doing, we will see that while there is some alignment between the waste reduction and diversion strategies offered in the campaign and those our minimal food waste producers enact, there are also significant points of difference that primarily manifest in an openness to being moved by food and the resulting capacity to underpin the development of the capacities to adapt to the uncertainties of food supply. Attention to these differences may assist in developing more nuanced approaches to food waste reduction capable of encouraging waste-reducing behaviours among more diverse populations and, as we will see, could also support the development of non-anthropocentric modes of being and doing that support broader resource-use reduction.

The trouble with food waste reduction campaigns

The Love Food Hate Waste strategy (LFHW) was developed and deployed in the UK by the not-for-profit group Waste and Resources Action Program (WRAP) in 2007. WRAP has worked in concert with a range of individuals, government agencies, businesses and community groups to promote food waste reduction. LFHW has also been employed internationally in New Zealand, Canada and

Australia. In Australia, the campaign exists in two states: managed in New South Wales by the NSW Government's Office of Environment and Heritage while being delivered by the Environmental Protection Authority; and, more recently, adopted in Victoria by the State Government with the campaign being delivered by Sustainability Victoria. LFHW tends to perpetuate a personalised, neoliberal-informed imperative for people to reduce their food waste. The focus is squarely on the individual householder's capacity to reduce waste if they follow the tips and tricks offered via the campaign's various websites and information resources.

LFHW asserts that its aim is to 'help you avoid food waste, save time and money, and reduce your environmental impact by planning better, shopping smarter and storing food effectively' (WRAP, 2017). These are, of course, shifts in practice that require a significant investment of time, often requiring not only a shift in beliefs but changes to ingrained bodily habits and adaptation to the different affective responses these might elicit. The key incentive encouraging people to engage in these shifts in the LFHW campaign centres on advocating the economic benefits of food waste reduction. However, the work with householders informing this text rarely identified financial drivers as being key to people's food waste minimisation behaviours. Furthermore, the previous chapter identified the need for differentiating between practices of thrift and frugality, as the former is unlikely to promote sustainable resource use. Indeed, not all forms of money saving will contribute to food waste reduction.

The limits of economic incentives and information deficits

Economic benefits are central to LFHW messages, most evident in the oft-cited monetary quantification of our food waste. On the UK homepage, the dominant message screams 'Saving you money, Saving your food' (WRAP, 2017), with the food waste reduction app that enables planning and tracking of household food waste flows being promoted as a 'money-saving app'. In NSW, at the time of writing, the campaign identified three key messages on its homepage; 'Save Money', 'Save Time', 'Eat Well'. The economic incentive is less overt in Victoria where more attention is given to the materiality of the foods and their qualities—namely taste, with the statement 'Don't let great taste go to waste' (Victoria State Government, 2017). However, the monetary incentive remains key to the messages designed to encourage behavioural change. The campaigns all employ aspects of a deficit model approach whereby the provision of new information and promises of financial benefits are assumed to support behavioural changes. As identified in the preceding chapter, these approaches are common in waste reduction initiatives, with Bulkeley and Gregson observing waste reduction initiatives are 'firmly tied to knowledge (awareness raising) and the hip-pocket nerve (financial incentives) as the primary barriers to/drivers of action' (2009, p. 934). However, a growing body of research on waste behaviours (Bulkeley & Gregson, 2009; Evans, 2012) demonstrates that 'there is potentially more than information and hard cash at stake here', instead pointing to the importance of social, physical, spatial and material concerns (Bulkeley & Gregson, 2009, p. 934).

The focus on information deficits in the LFHW campaign is identified by Evans, Welch, and Swaffield (2017) in their discussion of the website's recipe tool that is designed to help people use up leftovers or specific foods that are in danger of becoming waste. The assumption informing this, they suggest, is that people lack knowledge of what they could do with the foods and/or simply don't have the skills to prepare the foods languishing in their fridge or larder. However, Evans (2012, 2014) demonstrates that even when householders do have the skills and knowledge they may not draw on these due to broader domestic conditions such as not wanting to provide too much of one sort of food for health reasons or concerns about providing something all members of the household will eat.

While LFHW provides an excellent resource and some handy hints, campaigns that cast people as poorly informed, profligate consumers, are unable to adequately attend to the complexity of the relations within which systems of waste management operate in our homes and lives. They may also miss the opportunity to understand the strategies people currently employ to minimise waste and how these could be supported, enhanced and amplified. While the frameworks developed in LFHW may promote greater attention to the materiality of waste, the principal ways in which the campaign is rolled out fail to maximise the opportunity to use this to prompt a broader rethinking of anthropocentric thinking and practices which I contend is embedded in many waste avoidance practices. Instead, attention to the materiality in current campaigns tends to reinforce anthropocentric norms.

Perpetuating a nature/culture divide and fear of foods' becomings

In the LFHW campaign, and the broader environmental and social justice discourses related to food waste, food presents as something that we need to act on to avoid trouble. We are implored to love it while hating the waste that we might produce if we don't perform our love in the right way. It is a uni-directional relationship that, it tells us, will save us time and money. It can, of course, also work to alleviate any guilt people might be experiencing around their management of household food flows. The focus on the great taste of food, as emphasised in the Victorian version of LFHW, suggests a form of lively materiality of food, recognising its capacity to induce pleasure and incite responsive action. But by and large, food is represented in the LFHW campaigns as troublesome, particularly due to its ability to rapidly transition into something inedible that can pose a health risk as well as representing monetary loss. For LFHW, the troubles posed by food are to be averted through attention to four key sites: Planning (writing and adhering to menus and managing portion sizes), Shopping (writing a list and purchasing items with longest use-by dates), Cooking (using recipes for making use of leftovers and preserving) and Storage Tips (ensuring things are correctly stored, making use of freezers and rotating food through the fridge to avoid things getting lost). Even though the campaign promises its tips will save us time and money, the practices it encourages require hard work for some members of the households.

The householders in my research acknowledge that food waste avoidance does require labour or effort but this tends to simply be part of their habitual practices and any negative associations with this appear to be outweighed by the pleasure they express in relation to food. The attentive management of food flows and transitions of these householders is motivated by an excitement and valuing of foods materiality and their attunement to this, rather than a desire for control and fear of its lively potential to transition. This attunement is manifest in two key ways that I conceptualise as: moving and being moved by food (a form of receptive response-ability) and appreciation of both abundance and scarcity (the capacity to be flexible in response to food resources and to make do in the face of the inability to concretely plan for and control resources). Both of these, I argue, can be conceptualised as being enacted by the householders in my research through a playful approach guided by convivial dignity.

Moving and being moved by food

The householders in this study were all moved by food. Buying and growing food induced pleasure, excitement and connected them to 'something bigger' including the lives of growers, climate and their particular growing plots. Food was not merely a source of nutrition nor was the provision of weekly meals talked about as a chore, though time pressures were often cited as having an influence on particular shopping or eating habits. They enthused about food and were keen to talk about, and show, how its materiality moved them to cook certain dishes, gift goods and provide certain forms of nourishment for themselves and their family. Uncertainty, rather than concrete plans, defined most people's approach to food flows and this manifested in a high degree of flexibility in relation to the way foods would be consumed. This uncertainty encouraged attunement to vulnerabilities. Not knowing what food would be available for them to harvest from the garden or purchase from their preferred AFN due to seasonal variation, specific weather patterns and shifts in climate encouraged an ethos of adaptability and contentment with making do. This was not experienced as a form of going without but rather embraced as a necessarily responsive approach to the materiality of the food encountered.

My understanding of these viscerally experienced practices of 'making-do' involves an openness to being moved by matter and draws heavily on the notion of being a maker responsive to the materials being used. As we saw in Chapter 7, Carr and Gibson describe makers as those who work with materials 'within a realm of possibilities that are afforded by its particular properties' (Carr & Gibson, 2016, p. 303). Just as one might work with the grain and feel of wood to make something new from its form, the householders in this research worked with what was on offer to feed themselves and their families. Being moved by food regularly involved shopping without fixed lists, a lack of concrete meal plans, intentional cooking of excess (leftovers) to be reused for other meals, embodied engagement with the quality and freshness of food (rather than strict reliance on use-by dates) and an ongoing process of monitoring (including checking on availability from gardens) and responding to food so it would not transition to the waste stream. These practices do not map neatly onto the LFHW campaign approaches, though there are some key crossovers such as an emphasis on monitoring and moving or rotating food through fridges. In an effort to identify ways in which food waste reduction initiatives might be improved and possibly stretched to support broader practices of resource-use reduction, below I tease out the similarities and differences between the LFHW campaign and the practices of these food waste minimisers.

Food planning: flexible lists and responsive meals

List writing and meal planning are central to the food waste strategies advocated by LFHW and other food waste reduction initiatives. However, among the Canberra householders, a more flexible approach to food provisioning is evident, one that tended to incorporate responsive observations as typified by Mary's statement that she starts her food shopping by 'just checking out what's there'. While this may seem to be a typical feature of shopping habits, this degree of responsiveness to what was on offer was not practised in supermarkets by these same shoppers. Instead, supermarkets were described as places that were moved through as quickly as possible, where only essentials were purchased using a list that was, most often, strictly adhered to. People described their experience in the supermarket as feeling like they were a 'zombie' (Anja) or a 'bunny in the headlights' (Amelia) with others commenting that a list was not only an efficient way of navigating these forms of infrastructure but one which enabled them to escape without succumbing to various marketing and advertising ploys. These were not places they wanted to linger, nor were they sites that prompted responsive interactions with the goods on offer. Indeed, to pay greater attention was, for some, akin to leaving themselves open to succumbing to unneeded purchases such as 'two for one' offers.

The aim of the householders in this research was to get in and out of supermarkets as soon as possible with what had been pre-determined as needed. The different forms of navigation employed at supermarkets versus farmers' markets was exemplified in the comments of one participant, Kirsty:

> I do make lists [at the supermarket] and I'm fairly disciplined, we have budgetary restrictions so I'm not even tempted to buy extra chocolate or biscuits at the supermarket, I guess I'm old enough now that I've been through that phase. Usually I come home with what I go out to get. When I go to the markets that's not the case, I shop very seasonally and there's a couple of items I get every week like mushrooms and bread and fish but I very much choose the rest of what I buy on what's available. That's what I love about it, we're only eating peaches in season and the kids get really excited... they'll ask if it's stone fruit time and I love that.

And another, Alicia, who noted:

> I make a list to go to the supermarket for those essentials that you need to run the house, but I never make a list to go to the markets, or the health food shop, or anything like that. It's... the markets, there's sort of the basic things that I buy, but it also then depends on what's there. You know, something else might be in, something that... and I might see that it looks nice, oh well I might make this with it. Oh yeah, while I'm there then I might decide that tomorrow night we'll have this, or the next night, whatever fits in with what's happening in everyone's lifestyles.

A key driver of the main food waste reduction campaigns is the promotion of financial benefits of list writing and meal planning. However, in the above examples we can see that even when budgetary restrictions are in place, detailed planning is not used at the AFNs. This was common across the interviews. However, as Evans (2012, 2014) observes of the conventional shoppers in his research, a lack of concrete plans and lists does not mean that the shopping has no guiding structure. Most of the Canberrans had a sense of how much food their household would consume during the week and as they made purchases at the markets or farmers' market retail outlets they were usually thinking about the meals it could lead to. So, while the LFHW advice focuses on pre-shop planning to assist food waste reduction, these food provisioners had an embodied sense of household needs and developed plans for meals in response to what was available to them. This flexibility and adaptability, while leading to minimal waste, was driven by the uncertainty of what foods would be available rather than the use of a rigid pre-planned list of ingredients they could be confident of securing (as would be more likely to occur in a supermarket).

The pleasures of uncertainty for the food secure: attunement to the materiality of food

The experience of uncertainty—of not knowing exactly what will be on offer when—was experienced by the Canberra householders with a mixture of excitement and pleasure. This was seen to make food provisioning fun. There was delight expressed in their need to shop seasonally and to be flexible with ingredients. The mounting anticipation of seasonal food was always identified as a pleasurable challenge by these eaters, as we saw with Kirsty, the budget conscious householder above, and is further exemplified by another farmers' market shopper, Alicia, who talks about 'the fun of who pops up and what they've got and what they haven't got'. Indeed, for this participant going to the farmers' market is all about:

> ...the experience, because there's the same stalls there, but every now and then... like the apple store pops up every now and then, or come October

> the peaches and nectarines, like we sweat on the peaches and nectarines, and if you miss a market and they're there, everybody tells you about it.
>
> Because you wait for weeks, and weeks, and weeks, thinking they must come, they must come soon, they must come soon.

This responsiveness to the foods on offer prevents strict adherence to predetermined and stable patterns of consumption throughout the year. This requires flexibility in both what is eaten and how it is cooked or prepared in response to seasons and flows of availability related to weather, pests and disease. It requires the householders to become attuned to the myriad of factors that might make certain foods available and others not.

The specific food on offer and its characteristics directly affect household food habits and encourage heightened attunement to the relational more-than-human impacts on our food. This seems to encourage a non-economic valuing of the food itself. This value emanates from the food's very materiality as well as recognition of the relational entanglement of resources that are invested into its production, both human labour, care and time, as well as non-human inputs such as water and energy. This recognition, value, and pleasure in food can contribute to the development of more adaptable, responsive food practices: 'I never plan at the Farmers' Market, I just get what I like the look of and then I come home and then I go, "Oh, this is what I've got, what can I do with it"' (Gabriella). While in supermarkets, this form of responsiveness may be represented as impulse buying, in the AFNs, this attunement and responsiveness encourages the development of flexibility that can contribute to the enhancement of adaptable eating and cooking skills that may well be required in our uncertain futures. It is enacted in a playful manner capable of inducing great pleasure, as well as, at times, frustrations.

Rethinking an ordered kitchen: valuing responsiveness to food's materiality

Questions related to meal planning were often met by chuckles and statements about not being that organised. This lack of organisation, as indicated above, actually seems to refer to a non-normative notion of organisation. For those attuned to the material relationalities of food, an organised kitchen was not one that had rigid meal plans but one where the food available moved the householders to act in certain ways. So, rather than the householders' following strict eating schedules driven by particular placings of food, the materialities of the food themselves moved, or placed, the householders, enhancing their responsive capacities and flexibility. This was evident in the way meals were planned and the adaptive approaches to recipes as exemplified by Alicia:

> I try to follow the recipe... no, I don't plan the meals for the week. It just doesn't happen that way. I have a bit of a thought about what might happen, but that can turn upside down. I try and follow a recipe correctly the first time, but then I just... it morphs into whatever I want to do, and I never cook it the same twice. It just doesn't happen.

These householders viewed themselves as operating outside of the dictates of an organised kitchen and some felt they should become more systematic with their food management to impose more order (to fit an imagined ordeal of efficient household managements). But, by and large, this level of responsiveness appeared to be contributing to the production of minimal waste, far less than the oft-cited one-third of food purchased. It also seemed to enable these householders to develop and refine the skills required to adapt to unexpected changes to food needs throughout the week, such as when there were fewer or extra people to feed or when there was an unexpected night out. The responsiveness to food marked by the absence of concrete list writing and meal planning, and lack of reliance on specific recipes, seemed to assist waste reduction in their homes. These householders tended to design meals around the food available to them. They were responsive to the foods that they were growing, or those available at farmers' markets, in their food box or at the farmers' outlet. This is not simply a matter of eating seasonally but about playful tinkering with different foods, different recipes and different forms of eating. These skills in adaptability just may contribute to enhancing our capacities and affordances to adjust to our uncertain food futures.

Multi-pronged food waste reduction

Let me just clarify that the arguments made here do not seek to imply that list writing and meal planning are necessarily detrimental to food waste reduction. In fact, these are practices that may well be of great benefit to some households. In the householders encountered in this research, there was one participant, Lynn, who was a diligent list writer and meal planner, citing her lifestyle (part-time work, two young children in numerous activities) as requiring this level of organisation. For her, the meal plan was particularly effective at saving time and was also motivated by a desire to reduce her family's generation of food waste. If the items she needed were not available from the AFN she frequented she would source it elsewhere to ensure the meal plan could be adhered to. Sometimes, this was a result of not knowing what was in season or would be available. Though acquisition of this knowledge was starting to occur for her through her participation in AFN shopping, Lynn notes:

> I don't know what's in season, well I know the big things like tomatoes and mangos and very obvious ones, but other things... I'm learning a bit, but I figure [the farmers retail outlet] would source what's good, what's in. I'm relying on their judgment more.

In this cohort of participants, the lack of knowledge and attunement to food's seasonal and weather fluctuations was an important prompt for strict list writing and meal planning. However, participation in the AFN's was found to encourage the development of a more responsive and attuned approach:

> Yeah and sometimes from what I see in [the farmer's retail outlet], it'll be 'Oh beetroot's looking good, next week I'll do something with some'

The overriding concern for this householder, Lynn, however, was with not buying things that weren't needed. Her focus was on waste avoidance and these strategies enabled her to achieve this and they fitted in with her lifestyle. For others, the very vibrancy of the food encouraged responsive actions that prevented waste.

Making-do: the pleasure of dwindling supplies for food secure households

For the majority of these householders, responsiveness to the materiality of food itself was key to the way food entered into, moved through, and left their households. Their embodied and visceral delight in the food itself, 'its look, smell and taste', often meant that, after an AFN shop or garden harvest, big batch cooking was immediately carried out to preserve the food for the week ahead. For some, such as Gabriella, the 'dazzling' qualities of the food meant they often feasted straight after the market, with the food available in the house becoming less appealing and interesting as the week progressed:

> We usually eat really well on Saturday and Sunday 'cause especially if I've bought something like seafood, we'll often have like a cooked lunch' cause I'll sort of come home and go 'Oh, beauty, we've got that squid', and so you know we'll have a kind of three o'clock in the afternoon meal on a Saturday or a Sunday and then it kinds of peters out during the week.

Dwindling supplies throughout the week also invoked a level of excitement and interest in what food the market would yield the following week. Running low or out of fresh food (remembering these are not households in danger of being food insecure) was viewed positively and our eaters were mostly happy to make do until they could replenish the supply. This making do enhanced attention to the shifting states—the vitalities—of the foods. Once again, this challenges the positioning of modern minority world consumers as wantonly profligate. Instead, those in this research invested time and effort to respond to food's vitalities. This echoes the findings of Waitt and Phillips (2016) who identify practices of placing, rotating and assessing as key to how people manage food circulation through their fridge. They also note that keeping food visible is central to avoiding it becoming waste. This is reinforced in this research where participants observe that abundance—understood here as being more than can be consumed through everyday eating—works against effective food management and is avoided, as noted by Catherine:

> I don't like fridges that are packed with all these jars and things, I like a fridge that's got mainly fresh things. For a certain part of the week it's fairly empty and then it fills up again.

Kirsty, reflecting on her practices in a food diary, comments on the need to further adapt her habits:

> Fridge has too much stuff in it these days and I overlook some items, then they go off. Need to cut back a bit on purchases. This habit of stocking up that most of us have is a hard habit to break and a ridiculous one given our ready access to fresh produce these days. My cupboard does not reflect the fact that I can buy just about anything any time within five mins. drive. I am going to change that.

For most, there was a conscious effort to reduce and make do. As Alicia declared with pride, 'My fridge is rarely full, my pantry nowadays is rarely full and my freezer is rarely full'.

The enjoyment of the diminishing supply of food throughout the week encouraged practices of eating whatever was to hand which, again, induced an affect of joy through the playful tinkering with recipes it prompted. This is typified by one participant, Judy, who, when asked how she decides what to cook, stated:

> Okay. It probably depends, yeah, if I've got to the shops and what's in the fridge, because if there's stuff that I know I need to use, I'll just concoct something with those things before it starts to go, because I don't let that happen. And I suppose I get ideas from recipes, but I probably will look at a recipe and then modify it depending on (a) what I've got in my cupboard or fridge or whatever and also what I like and how I can add my little bit to it.

Ongoing monitoring and responsiveness to the transitions of food determined what was eaten, encouraging skills of adaptability and flexibility. For some, this was complicated by likes and dislikes of householders and fluctuations of movement in and out of homes. Planning meals was seen to be more difficult for those with families with children and more transient flows. As Jacquie says:

> So, I'm trying to reduce waste particularly with people being at home or not home it's very tricky with teenagers. Trying to plan and you either don't have enough or you've got heaps so you end up, you know, if I cook something and she decides that she's not going, she's going to be out then suddenly I'm either having to eat the same thing every day to eat it all up or I'm madly trying to eat and cook and things to use up stuff before it goes off.

Older couples or single person households often described their meals as boring, indicating they were quite content to eat the same thing for a number of days in a row or to have a variety of premade frozen soups as their main meal. However, regardless of the challenges, the vitality of the food and desires to avoid waste called the bodies of the primary food providers into action.

While strict meal plans and lists were mostly avoided, the one consistent mode of planning that occurred among these householders related to leftovers. Leftovers were deliberately produced and planned for in the shopping and cooking practices. It was common practice for extra serves to be made and stored away for lunch the following day or divided into portions and frozen for later. Unlike in Evans' work (2012, 2014), for these households, it was rare for leftover food to be stored in the fridge only to become waste later. This was a little trickier in share houses where some householders were away frequently and the others were never quite sure whose food was whose. There was also some waste associated with leftovers containing meat and rice that some participants felt raised food safety concerns, and the occasional loss of something that had been pushed to the back of the fridge.

While regular engagement with fresh food, moving it and being moved by it limited waste of these foods, it was a different matter for jars of preserved or processed foods where only a little was used at a time. The less overt presences and visceral engagement with the potential transitioning of these foods seemed to keep them from being actively managed the way that fresh food, or freshly prepared food, was. However, when these foods were encountered, the participants enacted responsive, embodied interactions to assess whether the food could still be eaten. Use-by dates provide an interesting point of discussion to explore these responses further.

Use-by dates, sensorial training and the impact of household demographics

For the householders in this research, use-by dates were typically used as a guide and, when shopping, people often sought out products in-date for the longest period (one mother referred to her son's 'long arms' as a key waste-reduction strategy for supermarket shopping as he would reach to the back of refrigerated cabinets to get goods with the longest use-by dates) as a means of preventing potential food wastage. However, these dates were rarely adhered to in a strict manner. Similarly to the findings of Waitt and Phillip's (2016) in their study exploring peoples use of fridges, the Canberra householders drew on multifarious means to make judgments about food safety and quality. They relied on embodied, visceral experiences—looking, smelling, touching and tasting. Being past the use-by date did not automatically assign something to the waste stream in most households. As exemplified by Judy, the use-by date was viewed more as an indicator rather than a golden rule:

> I smell it and maybe taste a little bit and see if it's okay and kind of try to use my senses a bit, and if it smells then I won't, yeah. Because often I think they can put something, the use-by date to be on the safe side and maybe it can last for another little bit.

Gabriella noted, 'I tend to completely ignore them you know? If it looks okay and smells okay, I'll eat'.

The willingness to view use-by dates as guides rather than as rigid rules was regularly linked to past food experiences and, therefore, confidence in the skills and capacities of the householder to make an informed judgment. For some, this involved growing up without use-by dates, for others it was a result of learning about food from parents and grandparents. Kirsty, our budget-conscious shopper from earlier, relied 'much more on [her] sense of smell and sight than use-by dates for sure', describing this as generational knowledge passed down to her through shared food encounters with family in her childhood. However, through the course of conversations for this research, she came to identify that this was also enhanced by her adult experiences with a local purveyor of fresh meat and through encounters with food in less sanitised and regulated environments:

> There was a shop owner ... when meat was out of date—and I don't know how sound it is but it certainly worked for him and it worked for us—he just throws it down and sometimes it almost goes black and he says it's going to be the tastiest meat. I trusted him and I cooked the meat and it was because it had tenderised but hadn't actually gone off. I don't know the science and that was one experience. I also worked in a kibbutz in Israel in the kitchen for 1000 people and a little old lady who ran the kitchen used to throw chickens in the soup that were green and I used to say to pull them out and she used to be very, very annoyed with me but no one ever got sick. I do throw chicken out because I think it smells but I've learned through experience. What's the worst that can happen in our culture? You can get salmonella, of course, but it would smell and so I rely on my senses.

While the discussions with the primary food provisioners in these households indicates an embodied and flexible approach to food marked by being open and responsive to its material qualities to assist in making decisions about food safety, there was significant variability in responses to use-by dates within households. Commonly, children and those who did not regularly provision or prepare food were more likely to adhere to use-by dates, as Sally recounts:

> ...my daughters will not drink milk, even if it smells absolutely fine, if it's past the use-by date. Whereas I don't have any... if it doesn't smell, it's fine to use, I'll make sauce with it or whatever. But they are very driven by that and I think that that's hugely... in fact you might be horrified, but tomato passata, well like puree that you buy in bottles, I'll open it and put it in the fridge, and then maybe I won't use it for a few days or a week or something like that a bit of mould will grow around the top, so I'll just remove the mould from the top and use what's in the bottom because the mould was not on the bottom (laughs). And in fact even tubs of say Philly cheese or that sort of stuff, if it's got mould on the top I'll just remove the mould and use the cheese that's underneath. My daughters are completely horrified and disgusted but you know, first of all probably if you ate the mould it wouldn't do you any harm it just might taste a bit funny or feel a bit funny. So, yes, I

> tend to be far more frugal about using up stuff than my daughters are, but they've been brought up in a different world.

While for those brought up in this 'different world' the food no longer had value past its use-by date, for their mother, the majority of the food continued to be edible if the bits that had gone bad were removed. This was prompted by a desire to be frugal and an acceptance that feeling a bit funny (if a poor choice was made) wasn't a significant issue. This fuelled this participant's scepticism about use-by dates, but others also recounted stories of use-by failures that reinforced the need to always draw on bodily senses to assess foods. This was exemplified by one story told by Mallory of a supermarket purchased item that was deemed to be off even though it was still within its use-by dates:

> ...we don't have that much [food] with a use-by date, but like I bought some tofu yesterday from bloody Woolworths—here you go—and opened it to cook with tonight, and I went, 'That's really off.' And I got someone else to try it, and she went, 'Yeah, that is off'. And Alice said, 'When are the...' because the tofu people [at the farmers' market] are on holiday, and Alice said, 'When are they coming back; I'm hanging out; they make the best tofu.' But it was off. It really was off. And it wasn't that we were being fussy, and the use-by date wasn't for another couple of weeks, it was under the use-by date.

Among these householders, multi-sensory engagement—observation, touch and smell—were all used to attune them to the transitions of the food. This was not particularly driven by food safety concern or economic loss but was much more about avoiding waste. The materiality of the goods and their becomings required this ongoing awareness and responsive behaviour. The characteristics that inspired the initial purchase of the food were often in a process of gradual loss throughout the week, and desires to enjoy these foods and prevent waste were paramount. This was marked by a sense of making do in response to the food on offer and its transitioning states. Moving food moved the bodies of these householders and encouraged engagement in playful tinkering with recipes and meals. This became particularly apparent when managing small-scale fluctuations of both abundance and scarcity.

Appreciating abundance and scarcity

The skills and capacities to manage both scarcity and abundance are central to being able to live in times of uncertainty, perhaps most viscerally evident in relation to a changing climate and resulting fluctuations in weather (Head, 2016). The people in this research exhibit these skills and capacities at the small-scale level of the household most overtly through their adaptability in response to the becomings-with of food. This ability to respond to scarcity and abundance appears to be enabled by an openness to being moved by the relational materialities of food. This form of responsiveness to the shifting multispecies entanglements of food—a manifestation of recognition of the inescapability of togetherness-in-relation—encourages movements, or placements—of food in particular ways that can reduce waste. Appreciation of the characteristics of food, the very liveliness of its materiality and responsiveness to this encourages forms of engagement marked by playful tinkering guided by convivial dignity.

These interactive encounters with avoidance of food waste can enable participation in what Haraway refers to as ontological choreographies that are capable of generating alternative ways of being and doing. A charge is created through these becomings-with-together that may well be able to motivate and support more sustainable living behaviours. In this research, this occurs for both those moved by shopping at AFNs and those producing food in their gardens. However, abundance and scarcity are often more dramatically and viscerally experienced by those directly involved in production, and so the following section focuses on these gardeners.

Playful tinkering with production: honing the capacities for multispecies adjustments

Food-producing gardening regularly facilitates experiences with abundance and scarcity that prompts gardeners to adapt their practices, ranging from what they grow through to how they will prepare their meals. Here, surplus is not spoken about in relation to economic value (such as the cost of managing over production as occurs in large scale food production) but is instead discussed in more multifarious ways from providing an opportunity to consolidate or build social connections via gifting through to the need to be inventive when inundated with seasonal gluts (zucchinis were a regularly cited example of this, even leading to the prize-winning pickle at the agricultural show discussed in Chapter 6). Throughout my fieldwork, I never left a garden without some sort of produce, from seeds to pickles. The gardens are places where fecundity is nurtured, encouraged and shared. But, while abundance is often welcome, it requires careful management to prevent it transforming into a burden. Unlike in larger scale agri-businesses where a flooded market and a lack of alternative distribution channels often leads to avoidance of the costs of harvesting, resulting in produce being left to rot in fields (likely to be quantified as a form of food loss), the small-scale gardener works hard to prevent any produce going to waste. This 'hard work' often manifests in playful tinkering where the relationalities of foods are reconfigured into other lives through gifting, via preserving, freezing and composting as well as via alterations to gardening practices.

The practices of managing abundance and scarcity for those involved in this research tend to occur through a mode of experimental play that draws on a tinkering approach (Mol, Moser & Pols, 2010). When these householders first started gardening, it was common for them to find themselves with far too much produce of one type and not enough of another. Through ongoing tinkering with the varieties and number planted, as well as the care provided (such as thinning of crops), they adjusted their practices to reach a level of production that worked for them and their plots or they adapted their eating habits and modes of food preservation to attune to the gardens' productivity. Often this involved the introduction of new infrastructure into their lives, most commonly in the form of a freezer. This was exemplified by a gardener, Greta, with the largest possible plot size in a community garden:

> . . .the first year when I had no idea of how much of anything I had to grow, and I had far too much. So, I had onions and I had sweet corn, and I had beans. I had tomatoes and potatoes, cucumbers, zucchini, capsicum. And come from midsummer on, I suddenly had this corner over here, just too much. And I started making soup. And initially I followed recipes, but very soon I just sort of thought, you know, I've got this, this and this. And I just made up lots of different soup concoctions and froze them. Then I had to buy another freezer.

Small-scale gardens are rarely sites where steady patterns of production can be assured. Playful tinkering is an ongoing, often daily process, enacted in response to the unpredictability and recalcitrance of weather, climates and presence of pests and disease. Abundance was not something that was seen to occur simply as a result of human actions but was produced through entangled relations of humans and nonhumans such as weather and climate which could lead to the unexpected growth and self-seeding of plants. It was common for people to note that they had failed to manage their crops properly by thinning out and discarding, something often expressed as 'just letting the plants get on with it' or to 'do what they want to do'. There was regularly a reluctance to 'manage' these plants in accordance with established gardening wisdom as exemplified by Gabriella's observation:

> . . .so even if it's something in my garden that came up wild so you know I haven't put any effort into it at all but it exists and it's edible and can be turned into something it kills me to do something else.

While for some this was part of the gardening ethos of permaculture, for others it was simply linked to the nature of a garden where the unexpected was always being encountered. Most often, gardeners worked around these unanticipated intrusions and adjusted accordingly.

The social pleasures afforded by abundance

The entangled human/more-than-human relations encountered in gardens required ongoing adjustments to plans. The gardeners engaged in this research aimed to produce only what they could manage, usually described as being enough to eat, share with family and preserve for the future. It was not an option to waste the food. Managing abundance was identified as a pleasurable challenge, one that enabled social connections to be forged or deepened. This included with neighbours, work colleagues, family and existing friends. Of course, as Bulkeley and Gregson (2009) have noted, social networks are key to avoiding surplus from entering the waste stream, yet not everyone has ready access to these networks. The necessity of forging such relations was seen to be an important way of managing garden produce with one newly arrived Canberran, Catherine, noting:

> Yeah. The fellow over the fence hands over tomatoes and zucchinis when he's got a glut of them; we hand over what we've got. We're developing that but that happened a lot [where we previously loved] because we all shared what we had too much of.

The role of the social networks became most obvious when interviews were carried out with households who knew each other (as occurred once in this study) and lived in close proximity. As one interview got underway, it soon became apparent that there was batch of feijoa gin fermenting in the corner. The feijoas were also being preserved and eaten fresh. It was when yet more feijoas were encountered in another household that the gifting cycle became apparent. These were households that regularly helped manage the other's surplus, as Mallory indicates:

> ...with friends, say they've got a lot of feijoas or a lot of tomatoes, then sometimes you sort of swap and go 'Well, I'll preserve them and then I'll give you something back'.

Even when there was no reciprocal swapping occurring, the gifting of abundance in the form of fresh produce was seen to be a way of nurturing families and also as a means connecting people with food flows and the process of production. This was about sharing knowledge and experiences—a stretching of the notion of the social as interactions among bounded entities to relational flows—through the material form of the food. As another gardener stated:

> So, what I'm saying is you'd have to scratch very hard to find all these traditions of growing things and people, so I guess most people, some people are probably quite divorced from it or have always been urban people, but other people have a slightly more rural background or more of different cultures that are more engaged with their food would be more into it and so I think people would benefit from people who grow too much then give it to their friends and there's that whole exchange thing which is a good thing.

While abundance in waste research might be viewed as problematic, within the garden and for AFN shoppers, it also has the capacity to induce immense pleasure, potential future sustenance and highlight human and more-than-human entanglements. Waste here is not merely excess, an example of a society focused on over production and over consumption. In these sites, responsiveness to the uncertainty of abundance allows for the practising—the playing out—and sharing of the skills and capacities, as well as awareness of the affordances, needed to adapt to excess to avoid waste. Indeed, abundance in the garden can be a thrill, as Warren notes:

> Leafy greens are really easy to grow and I eat them a lot so it's nice to just have kale growing in the garden and tomatoes as well, there's kind of a fun thing about you know the like obscene quantities of tomatoes that get produced and then we made 12 kilos of green tomato pickle with all of the left over green tomatoes, and you know raspberries growing in the backyard, like I think it's nice to have that real connection with the food and know where it's come from and the effort that it requires and, you know, I'm a lazy gardener so I'm never going to grow carrots and potatoes. It's too much effort but you know spinach I can grow in the garden and it just grows and that's perfect.

The pleasure of preserves and the capacity to provide ongoing sustenance offered by such foods encourages human labour. While this participant identifies as a 'lazy gardener', he willingly exerts effort to preserve the foods for later; they call him into action.

In the garden, the desire to avoid waste leads to people inventing, seeking out and sharing recipes in response to what has produced well. This was regularly conceptualised, not as work but as forms of play, as Marco observes:

> We have been doing the tomato sauce; we have never bought tomato sauce in the shops. Every year we make a couple of hundred bottles for all the family.
>
> It's a sort of amusement, when we do it you know it takes one weekend and we get my daughter's friend, usually they come because they want a few bottles.
>
> So, we stay all together cracking jokes and we have a barbeque and then… Yeah, it's a good party.

These experiences of abundance were considered to be a hallmark of food producing gardens and thus a playful approach to the use of produce is always being tinkered with to avoid waste. As Melissa points out:

> The idea is to work out different recipes just to be able to consume it all so you don't have the wastage. The idea of having an overload of things is to store it for use later in the year when there's less things around that you can grow so we've got produce all year round really then.

Joyful adventures with abundance

Experiences with abundance afford, and in fact necessitate, playing with modes of preservation and waste avoidance. While this is evidence of a frugal mindset where the value of the materiality of food drives efforts to avoid waste, it is not one that requires strict, regimented planning and activity. It tends to ignite practices that the Canberrans delight in. They are driven by a desire to make use of and share the 'beautiful' produce and the more innovative, the better. This was certainly true for one gardener, Ken, whose food production had significantly altered the style of food he consumes and developed in him a vociferous appetite for inventive dishes. After sharing with me a range of homemade pickles under the trees in his community garden patch at the end of a particularly hot summer's day, he offered me a gloriously purple preserved egg:

> I had all this beetroot and I thought well, what can I do with beetroot? And I said beetroot recipes [into a search engine]. And the first one to pop up was beetroot eggs. So I modified it a bit. Played. And the beauty of it is that unlike pickled eggs, you can—well you have to wait two to four weeks depending on what recipe you use—beetroot eggs, which is still a pickling process. They're edible in two days. And you just keep them refrigerated.

This playful attitude tended to dominate his food encounters with the garden produce prompting changes to his eating habits, in particular, increasing his consumption of preserved foods. This was a direct response to the abundance generated in his community garden where seed and seedlings were freely shared among plot holders and one might not always know what they were planting nor what or how much it would produce. This encouraged innovative, responsive approaches:

> It's a little more adventurous, but at the same time you feel quite smart about yourself—or chuffed—that, well you know, I need something different with a bit of bite to it. I know! I've got buttered breadcrumbs down… butter… bread and butter whatever it is cucumber, or a nice sharp pickled onion… And you can go and get it. You don't have to actually worry about going to the shop to get it.

While abundance has the capacity to elicit an affect of joy and encourage playful tinkering, its management does require certain infrastructure, social networks, adaptability and a desire to avoid waste, regularly expressed as frugality. With the right affordances, experiences with abundance of foods in gardens and in response to seasonal gluts via AFNs can be generative of creative, resilient, playful practices enacted in response to recognition of the necessity of our togetherness-in-relation which enable the development of skills and capacities attuned to uncertain futures.

Responding to scarcity

Scarcity is commonly linked to an ethos of making do. While a simplistic reading of abundance suggests that actors can make a choice to waste or not (though in this research the choice to waste was not considered to be a legitimate option for the participants and was studiously avoided), scarcity removes any appearance of choice. But, of course, one must have particular skills, affordances and capacities to make do. It doesn't just happen in response to lack. In the oft-cited examples of food scarcity in the minority world such as rationing in wartimes and the planting of victory gardens, those engaging in these endeavours often had skills (how to garden, preserve foods and fix/repair things), or contact with people able to readily share these that made the endeavours successful. These were times when goods were not so easily replaced and the global food system was just gathering strength.

The modern mega city has fuelled distancing and disconnection between food sources and consumption largely obfuscating evidence of the seasonal cycles and shifts of abundance and scarcity. Global agri-business has made produce available year-round so that scarcity is primarily experienced through higher costs. While people might go without certain foods for a short period of time, there is the assumption that things will return to normal in the not-too-distant future. This provides little incentive to develop the skills necessary to adapt to change. As outlined earlier, mapping this terrain is not an attempt to cast modern householders as profligate, unthinking consumers. My aim, instead, is to highlight how contemporary minority world material contexts mean that for those who have access to a comfortable life, making do is primarily experienced as a temporary state of waiting for normalcy to return. This temporary state limits its potential to support development of the skills and capacities to adapt, fix and/or repurpose materials. Of course, the uncertain futures induced, in particular by climate change, raise doubts about the potential for certain modes of living, including what we eat, to continue in accordance with familiar trends.

I must be careful here to not over generalise. This book attends to those who do demonstrate skills and capacities to adapt and explores how these might be expanded and extended to others. Scarcity in the modern minority world may not necessarily reduce consumption nor invoke the skills of adaptability and flexibility that may have been previously needed. Yet, in the gardens of food producers and those engaging with food fluctuations in AFNs, embodied, visceral experiences with both abundance and scarcity coupled with the carving out of space, time and development of the social connections to play with ways of attending to these can encourage and support more frugal, less wasteful lifestyles. For the gardeners, this playful tinkering in relation to scarcity is quite often a result of crop 'failures'—things not growing according to plan as a result of climatic conditions, pests, inadequate soil nutrients or poor human care. When this happens, the gardeners adapt, they grow something they hadn't intended to or wouldn't normally eat, something that suits their current soil profile, the season and location.

Playful adaptations in the garden: 'lively' frugality

Playful tinkering marked by adaptation was particularly common in community gardens where people often planted the extra seeds of fellow plot holders and then shared recipes for the unplanned-for harvest. The necessary alteration to plans made in response to crop failures were experienced as frustrations but were also talked about as being a 'bit of fun', part of the 'trial and error' of gardening and thus part of the gardeners' recognition of the limits of human control in these spaces. After all, as was frequently noted, there is always something that will grow. One participant, Pamela, who was new to food producing gardening observed, '[y]ou can't be too precious about some things and I think the land and the soil, sort of, they always come to the fore, to the party'. Indeed, the gardeners regularly talked about how 'nothing goes to waste' (Nina) in the garden, from the eating of bushy carrot tops from a crop that failed to develop roots through to the reuse and recycling of materials for garden bed structures and the compost and manures brought in from horse shows and agricultural shows. A frugal approach was consistently evident amongst the participants. While rarely linked to money, this was always linked to the very materiality of the things with which they engaged. Nonhuman 'things' were inevitably seen as having 'life' left in them, or being 'worth the effort' and able to be creatively repurposed to nourish and support humans and nonhumans alike.

While many of the gardeners were attuned to the realities of climate change, they rarely linked their gardening choices directly to these or broader environmental concerns. Instead, engagement with and extension of the life of material goods and adjusting, playful experimental practices were largely seen to be the right thing to do. They were key to a non-wasteful 'good' life. Even in the community gardens where organic gardening principles had to be followed, most people were at pains to tell me they hadn't sought out the plots for their organic principles. A number of participants wished to make it clear that they weren't 'hippies' or 'zealots', but they did find that gardening in these spaces and their experiences with abundance and scarcity encouraged awareness of environmental issues. As Bill noted, 'I guess that we're a lot more attuned to the importance of the environment, it's not that we're missionaries in that sense or activists but we're certainly a lot more conscious'.

While the fluctuations of abundance and scarcity are perhaps less keenly experienced by AFN shoppers that aren't also food gardeners, their responsiveness to availability did provide points of connections with these ideas. The cyclical nature of seasons, and the flows of availability related to weather, pests and disease, impacted directly on most householders' food habits and heightened their attunement to these more-than-human impacts on their food consumption. Engagement with these fluctuations may well provide a useful training ground for the development of skills and capacities attuned to the uncertainty of our food futures.

Conclusion

Throughout this chapter, we have seen that engagement with food waste and efforts to avoid foodstuffs entering the waste stream have prompted the gardeners and AFN shoppers with whom we have dwelled to enact a series of shifting, embodied practices responsive to the materiality of food. This responsiveness led to them both moving (circulating it through fridges, transforming it into other forms and gifting within their social networks) and being moved by food. The modes of food provisioning they engaged in also prompted encounters with abundance and scarcity that afforded opportunities to engage in playful tinkering opening up opportunities to develop the skills and capacities needed to adapt to the uncertainty of food flows to avoid waste. Through these interactions, recognition of the interdependencies of all life and our mutual vulnerabilities were brought to the fore. These involved multisensorial and multimaterial encounters including manifestations in temporal and financial vulnerabilities as well as concerns about not having enough to eat. By and large, however, the participants' approaches to food waste present as variants of frugality where waste was often cast as an enemy that had to be avoided. These frugal practices tended to be driven by a valuing of the materiality of the food itself and the relations through which it congeals and transitions.

Through paying close attention to these relations I have attempted to reinforce the need to broaden our tool-kit of material-semiotic understandings and approaches to food waste minimisation beyond those commonly manifesting in food waste reduction campaigns. Key to this, I contend, is the need to push at the boundaries of the social relations of waste and of the very limits of what we include in our concerns for the social, once again suggesting the need to stretch this out to incorporate a multispecies approach but also to move beyond bounded entities to incorporate relational flows and their effects and affects.

While waste relations are commonly informed by neoliberal forms of individual responsibility, attunement to food transitions can support the cultivation of more multifarious and generative modes of togetherness. These practices involve exposing ourselves to elements of risk as they produce encounters with vulnerabilities. However, vulnerability is understood through this text as not only constituting exposure to danger but, when enacted through play, also as enabling engagements marked by openness to both surprise and receptivity (Green & Ginn, 2014, p. 152). Recognition of the mutual vulnerabilities of all life could lay the groundwork for the promotion of more sustainable urban living. In this book, this generative potential is captured through the notion of convivial dignity which is used to provide a framework through which to narrativise the possibility, and lived realities, of non-anthropocentic practices and modes of being that may be capable of supporting more sustainable futures.

The following chapter builds on this identification of the capacity for waste relations to support attunement to mutual vulnerabilities and seizes on the importance of expanded conceptions of, and concerns for, the social by focusing our attention on composting and the nonhumans that assist with these forms of organic waste-management. While in this chapter we have dwelled with householders enacting attuned practices of waste avoidance through careful management and prevention of foods unintended transitioning into forms perceived as being inedible, in the slippery, multispecies microworlds of compost, these relational transformations of matter are not only welcomed but actively encouraged. The compost heap is shown as a site ripe with encounters with strange-strangers that highlights the necessity of our togetherness-in-relation and the potential for playful tinkering modes of engagements with alterity to expose the myth of human exceptionalism and encourage reconfiguration of practices attuned to the mutual interdependencies that enliven convivial dignity. Entering into these multispecies and multi-material relations involves risk, but the affective force of embodied, sensorial, visceral interactions with food can encourage a playful response, often driven by an affect of pleasure accompanied with a drive to make do.

References

Bulkeley, H., & Gregson, N. (2009). Crossing the threshold: Municipal waste policy and household waste generation. *Environment and Planning A*, *41*(4), 929–945.

Carr, C., & Gibson, C. (2016). Geographies of making: Rethinking materials and skills for volatile futures. *Progress in Human Geography*, *40*(3), 297–315.

Evans, D. (2012). Binning, gifting and recovery: The conduits of disposal in household food consumption. *Environment and Planning D: Society and Space*, *30*(6), 1123–1137.

Evans, D. (2014). *Food waste: Home consumption, material culture and everyday life*. London and New York: Bloomsbury.

Evans, D., Welch, D., & Swaffield, J. (2017). Constructing and mobilizing 'the consumer': Responsibility, consumption and the politics of sustainability. *Environment and Planning A*, *49*(6), 1–17.

Green, K., & Ginn, F. (2014). The smell of selfless love: Sharing vulnerability with bees in alternative apiculture. *Environmental Humanities*, *4*(1), 149–170.

Head, L. (2016). *Hope and grief in the anthropocene: Re-conceptualising human–nature relations*. Oxon and New York: Routledge.

Mol, A., Moser, I., & Pols, J. (Eds.). (2010). *Care in practice: On tinkering in clinics, homes and farms*. Bielefeld, Germany: Transcript Verlag.

Victoria State Government. (2017). *Love food hate waste*. Retrieved from http://www.lovefoodhatewaste.vic.gov.au/

Waitt, G., & Phillips, C. (2016). Food waste and domestic refrigeration: A visceral and material approach. *Social & Cultural Geography*, *17*(3), 359–379.

WRAP. (2017). *Love food hate waste*. Retrieved from https://www.lovefoodhatewaste.com/

9 Composting in the home

Introduction

In Chapter 3 I suggested that play can be understood as a form of composting, a mode of encounter that can transform the ways of being and doing of the entities 'in play' through decomposition. Here, I attempt to show that the reverse is also true, composting itself is a form of play that highlights the transformational potential of mutual interdependencies. Such play can induce affects not only of joy and belonging induced by attachments but of a much more putrid, confusing kind where detachment is desired while the very boundedness of the entities involved may be put under erasure. The death and decay of decomposition requires forms of togetherness or 'mutual inclusion' (Massumi, 2014, p. 6) that are produced through a 'thoroughgoing spatio-ontological reassembling' in which life becomes 'a vector of relation and recombination' (Brice, 2014a, p. 180). The experimental play of compost is generative precisely because engagement in relations with unknown and ultimately unknowable entities exposes mutual vulnerabilities and the limits of bounded forms.

Like digestion, these are sites where humans cannot do the work alone. However, while the human/more-than-human entanglements that shape digestion typically form without conscious human effort, in compost heaps and the bodies of worms and chickens, humans actively enter into these risky relations by inviting unknowable others into their domestic spaces. The resulting encounters with the inescapability of togetherness-in-relation are never simply harmonious. They are fraught with invasions of undesirable species: slaters that help with decay also feed on young seedlings; rats and mice that love to feed on the chickens' leftovers and enjoy the warmth and nourishment of compost bins; and the potential health-threatening pathogens lurking in the decaying matter. But, here in these tense multispecies landscapes, the limits of human control are exposed and, through play enacted with the cultivation of responsive and inescapable togetherness with more-than-humans and shifting relations guided by forms of convivial dignity, reconfigurations of anthropocentric modes of being and doing attuned to uncertainty are generated and enacted.

As we have seen, waste is only ever moved along. It never simply disappears. These processes of becoming ensure material goods remain present even in their

DOI: 10.4324/9780429424502-9

absence. For many in this research, home food waste management systems involving the guts of chickens, worms or in compost heaps and bokashi buckets keep these absences present through transitions capable of attuning humans to our entangled interdependencies. Here, the vitalities of these entanglements are encountered in visceral and embodied ways. The processes of decay and repurposing of food waste exceed the materiality of the what-was-once food itself and 'present absence' is experienced as a state of entangled becomings-with.

In the previous chapter, I discussed the ways in which the characteristics of food and its transitions (which are bound up with human actions such as methods of storage, as well as nonhuman actions, such as temperature and microorganisms, and the interactions amongst these human and more-than-human elements), as experienced in relation to the pleasure gained from food, can be understood as productive prompts for action, encouraging response-ability to the vitalities of food. However, the dominant messages related to food waste, as exemplified in the LFHW campaign, instead cast these transitions as troublesome and something that necessitates managerial efficiencies in order to avoid waste. In compost heaps, worm farms and bokashi buckets these transformational becomings-with are desired and encouraged. These contact zones lead to ongoing encounters with shifting forms that exceed materialities themselves and are only enabled through matter-in-relation. These modes of encounters may be generative of modes of being and doing that challenge dominant anthropocentric narratives through their embrace of the uncertainty and unknowability of these sites of non-bounded, entangled becomings.

The propositional nature of compost

Composting for many in this research was simply seen as being the 'right thing to do' and often something that had always been done in their extended families. As we saw in the preceding chapter, for some participants this manifests in dramatic physical and emotional reactions to the placement of excess food into mainstream waste. However, despite these ingrained bodily habits, the propositional nature of compost—its capacity to remake waste into something new—was consistently spoken about with excitement and a degree of awe. This 'awe' has some resonance with Jane Bennett's conception of 'enchantment', but is much less temporally and cognitively bounded. For Bennett,

> . . .enchantment entails a state of wonder, and one of the distinctions of this state is the temporary suspension of chronological time and bodily movement. To be enchanted, then, is to participate in a momentarily immobilizing encounter; it is to be transfixed, spellbound.
>
> (Bennett, 2001, p. 5)

Enchantment involves heightened sensory activity in which '[t]he world comes alive as a collection of singularities' (Bennett, 2001, p. 5). However, the excitement and awe identified by the composters in this research is driven more by recognition of relational entanglements rather than the characteristics of the bounded entities involved. It is the very assemblage nature of the propositional potential of compost that is beheld here. The wonder of, and at, compost is more dirty, earthy and co-constitutive than seems to be the case in enchantment. It is also not necessarily about finding points of attachment to support more ethical engagements with nonhumans but about the possibilities of detachment and other relational modes that manifest in the recognition of having to live together. This is not to say that enchantment does not offer a potentially productive prompt for new modes of being and doing in the Anthropocene, but it is a concept that does not map directly onto the practices, and ways of speaking about these, of the participants in this research. The stories told, and played out, here are suggestive of a more relational, propositional notion of play.

The propositional nature of compost was regularly spoken about as a means of alleviating guilt and anxiety related to the production of food waste rather than as a source of enchantment. However, it is important to point out that these efforts to off-set concerns were not considered to constitute an excuse for being wasteful. Food waste was still something participants actively worked to avoid as typified by one householder, Kirsty, who noted:

> There's virtually no food waste that goes into our bin, I would be surprised if any does. We've got a worm farm and chooks and three compost bins so everything leftover goes into those in some way but we're not very wasteful people. The children have to eat their meal, I don't overcook. We come from quite a frugal family tree so we save a little bit of food, I use leftovers the next day and we don't have very much waste... There's really no waste goes in our bins.

While the capacity for food waste to transform into something that could be productive—a resource if you will—was valued by composting households, people were still troubled by the loss of food's intended purpose to nourish human bodies as exemplified by Gabriella:

> It's one of those things about waste, you know. And I'm pretty relaxed about how long I'll leave it before I still eat it. I'll leave stuff; I'll leave a casserole sitting on the stove in this weather for like four days, just keep eating it. In the fridge, a couple of weeks. You know, if it doesn't really smell or look really dodgy I'll eat it. So, yeah, no, I really, and as a last resort it will go to the chooks. But I tend not to throw away, it's a bit like, you know, anything edible I think should be eaten, whether it's off a tree or whether it's something I've cooked. Particularly something I've cooked because even more energy's gone into it. Yeah, I'm probably a bit almost obsessive about throwing away anything that could be eaten.

Concerns about human bodies and the valuing of the resources that contribute to the production of food were also extended to those in distant places. Amelia, whom we met in Chapter 5, feeds surplus food to chickens which she spoke about as a means of 'easing' rather than eradicating her 'guilty conscience' about the 'starving children in Africa'. In these snippets of peoples' stories, we can identify the ongoing presence of absence that, while repurposed, continues to haunt and disturb.

More than bounded entities: the potential of matter-in-relation

Compost is a complex beast or a complex site of multiple beasts that all-together compose, while simultaneously decomposing, what compost is and can be. It is the inability to single out bounded entities as they shift and transform in relation with multiple others generating heat, pH shifts and other conditions that contribute to their ongoing reshaping and reworking that is the hallmark of compost's propositional potential. Or perhaps, more precisely, it is about the observation and sensorial engagement with the shifts in the form and breakdown of bounded entities and attention to how their becomings are induced by multifarious others. Food waste is ripe with potential to become something else, demonstrating the multiplicity of shifting lives and forms of matter-in-relation present in a garden.

The propositional nature of compost is evident not only in its formation, but in its potential uses. These sites of becoming were celebrated due to their potentially productive impacts. One community gardener, Michael, noted that 'mountains' of compost was 'every gardener's fantasy' because of its capacity (through the labour of micro-organisms and worms) to turn 'dead' matter into a form capable of increasing food productivity 'threefold'. The potential of 'dead' food waste to interact with other elements in the garden to forge new assemblages encouraged recognition of a collective vibrant matter. While the elements in isolation were considered inanimate, together they interacted to not only contribute to the production of new life (through their use on the garden) but to become lively multispecies aggregations. This potential vitality of food waste was seen to be an enlivening force squandered in conventional waste streams, as one householder, Catherine, observes:

> I wouldn't throw it in the bin, it would just go into landfill. It can go in the garden, it's feeding the garden so it's the rather passionate person in me that says... that whole micro diversity thing, things moving in the garden. I love that idea of helping to aerate the soil, the synergy of it all.

The assemblage nature of compost contributed to supporting recognition of 'things moving in the garden' interacting with and transforming states drawing attention to the 'multispecies landscapes' (Tsing, 2014a, p. 35) without which 'we cannot survive'. Indeed, as Tsing writes, 'We become who we are through multispecies aggregations' (Tsing, 2014b, pp. 229–230) and interactions with compost for these participants, enhanced attunement to its multispecies, as well as transformative relational nature, enlivening their embodied experiences of human vulnerabilities, namely their reliance on more-than-humans and the limits of human control.

Multispecies becomings-with

Attunement to the multispecies aggregations of compost were found to not only contribute to the production of new life (through its use on the garden) but also prompted alterations to human practices and provided a foundation for a reconfiguration of forms of human/more-than-human relations where positive, excited feelings were expressed about the 'rich and wormy' 'smelly stuff' that was able to nurture and produce 'clean, healthy food' to sustain human bodies. For some, this prompted shifts in bodily practices and daily habits. Bill, a retired townhouse dweller with a community garden plot stated:

> I think it's all linked in together. You use the compost out here. I mean, we would have thrown it out, now we have a garden to use it.

For him, the resulting encounters with more-than-humans prompted attunement to bigger picture environmental concerns (speaking specifically about the need to conserve water, a direct result of being interviewed during a period of drought). Speaking of his compost-rich soil he noted:

> It's great to get down and dig and feel the consistency of it [the soil] now and the worms in it. It's something that really strikes you. I never thought that I'd be quite turned on by putting my hand in the soil and looking at it. I guess the older you get different things appeal.

Putting compost to use in the garden and observing, feeling, smelling and even tasting the soil (Puig de la Bellacasa, 2017) emphasises the capacity for these multispecies aggregations to produce newness. Often these intimate encounters laid the foundation for new forms of attachments. Another community gardener, Karl, who had worked his plot for 20 years described himself as being so intimately engaged with his garden that he had come to know 'every grain of soil... personally'. He talked about himself as 'acting' on the soil through the addition of compost but this was coupled with careful observation of its impacts (such as its water holding capacity) and responsive shifts in his actions. Another permaculture gardener, Dinah, spoke of her compost recipes and how when she added it to the soil it then became capable of producing 'vitamins and minerals' for human consumption. However, the propositional capacity of compost is not reliant on the formation of intimate, harmonious attachments. It can also be realised through forms of detachment, or more nuanced forms of togetherness or 'mutual inclusion' (Massumi, 2014). Compost necessitates engagement with alterity. Not being able to know the other is a risk, exposing players to mutual vulnerabilities, but in so doing allowing the play of possibilities to unfold. Here we can see that compost draws attention to the inescapability of our togetherness-in-relation.

Compost as risky togetherness-in-relation: beyond attachment and detachment

Mutual vulnerabilities, understood throughout this text as presenting a 'condition of receptivity' (Green & Ginn, 2014, p. 152), expose our multispecies aggregations and, when recognised, have the potential to prompt significant shifts in practice, narratives and thought. These encounters with alterity and the limits of human control are risky, and this induces interactions marked by forms of playful tinkering. This does not necessarily lead to the formation of attachments—not even the risky ones Latour (2004) speaks of. But nor does it necessarily induce 'dreams of detachment' (Ginn, 2014, p. 541). While aspects of both of these may be experienced, following Abrahamsson & Bertoni, my concern here eschews a strict focus on the nature of relations (attachment or detachment) to concentrate instead on what 'this relation does, what it enacts' (Abrahamsson & Bertoni, 2014, p. 145). What we see in peoples' engagement with the multispecies aggregation of compost is a form of togetherness where 'strange strangers' (Morton, 2010) are encountered in the process of transformation. This togetherness of unknown and unknowable strangers can be constituted by interactions among individual, bounded entities, but also by 'flows of energy and materials' (Brice, 2014b, p. 186). Compost is a site where bounded bodies are exceeded—they break down—and alterity in relation is encountered. To enter into these relations is to engage in risk by decentering normative understandings of both self and matter by recognising the inability of us to ever be alone, and thus, the inevitability of our togetherness-in-relation.

However, while excitement at the potential of compost permeated householder narratives, recognition of togetherness-in-relation does not produce 'cozy' relations (Rose, Cooke, & van Dooren, 2011). Indeed, there are many awkward encounters with 'monstrous' (Ginn, 2014) others and recomposed relations in these sites. These tensions are evident in the stories told by those who participated in this research: here notions of mutually constitutive togetherness are hinted at, and certainly evident in bodily practices, but the narratives available seem poorly equipped to represent these experiences that unsettle dominant anthropocentric norms. These narrative limitations may also be encountered because compost emerges in practice, it is a 'doing' with and of matter, and as Law and Mol write 'while material politics may well involve words, it is not discursive in kind' (2006, p. 17). We shall return to these semiotic limitations shortly where I again posit convivial dignity as one means of attending to this, but for now I want to flesh out these encounters a little more to explore how the togetherness of compost plays out.

Playful tinkering with togetherness-in-alterity

Tinkering seemed to be key to the management of compost for most participants. The forms the tinkering take, can be seen to constitute playful experimentation.

The aim of these ongoing, responsive adjustments was to make conditions ripe for unknowable others to enter the relational entanglements and for new relations to form. To attempt to do this, there is a need to attune to their needs, likes and dislikes. As I suggested in Chapter 2, this can be conceptualised as attuning to 'strange strangers' (Morton, 2010). Morton's notion of the 'strange stranger' positions humans and non-humans alike as unknown and unknowable. The aim of ethical ecological engagement does not have to be focused on getting to 'know' the other or to form attachments. There will always be an unknowable, unassimilable aspect of 'us' and 'them'. The challenge then is to enact togetherness with alterity. Morton posits the notion of 'the ecological thought' to capture this which, as he writes:

> ...imagines interconnectedness, which I call the mesh. Who or what is interconnected with what or with whom? The mesh of interconnected things is vast, perhaps immeasurably so. Each entity in the mesh looks strange. Nothing exists all by itself, and so nothing is fully 'itself'.... Our encounter with other beings becomes profound. They are strange, even intrinsically strange. Getting to know them makes them stranger. When we talk about life forms, we're talking about strange strangers. The ecological thought imagines a multitude of entangled strange strangers.
>
> (Morton, 2010, p. 15)

These ideas jettison a human-centred understanding of connectedness, instead 'at each node of the network there is a radical gap. Our encounter with the network at any point is with an irreducible alterity' (Morton, 2010, p. 76). Here, Morton challenges notions of anthropocentrism and anthropomorphism by focusing on a need to 'respect' the 'alterity of the strange stranger', thus removing the capacity of humans to inscribe the world with static meaning (Morton, 2010, p. 76). This is akin to my notion of togetherness-in-relation that I attempt to narrativise through the framework of convivial dignity. However, while Morton focuses on the aesthetic world of literature and art, I delve into an altogether darker world: one of rotting flesh where encounters among strange strangers emphasise the vitality of matter-in-relation while imploding any privilege ascribed to bounded entities. Thus, it is not only the alterity of identifiable (even when unknowable) matter that convivial dignity attempts to enliven but the relational flows that produce it and us and are generative of transformations.

Shifting relations: multisensorial adjusting with the compost heap

Composting was integral to the practices of the food-producing gardeners encountered in this research. Many would say 'Of course I compost', a statement often quickly followed up with, 'but I don't do it well'. Others told me stories and showed me intricate composting systems that might include chickens, worms or fungi (lactobacillus). Often these were being 'trialled' in accordance with a pattern of regular, playful tinkering with ways to manage waste and produce a useful soil conditioner, though compost was never thought about as just an additive. Commonly, it was the process of cultivating the compost and the daily rituals associated with this that people shared with me. No one ever felt like they got it 'just right'. There was always something else to do, to learn or to try in the compost heap. This uncertainty encouraged the composters to continually adapt practices in response to material presences in the heaps. Time was invested in attuning to the compost's needs so it could meet their needs. These people were intimately aware that they could not do it alone, yet the workings of the compost and the flows of life and metabolic patterns within it remained difficult to know.

It is rare to talk to someone about compost without them wanting to show it to you. The showing invariably involves digging about and smelling it. I was regularly implored to stick my hand into garden beds the compost had been used on so I could see and feel its benefits; the improved moisture retention of the light friable soil and the tickle of thick, juicy worms so abundant they couldn't escape my touch. Composting, then, is very much about sensorial engagement as a means of attuning and tinkering. It is kinaesthetic in nature. The stories told are about success and failures. The failures are always attributed to the human components, occasionally aided by 'poorly designed' technology such as small worm farms that were seen by some to be too difficult to manage in the extremes of Canberra's subzero winters and blisteringly hot summers.

The composters assign to themselves the responsibility of getting the conditions right for the decomposers to carry out their task of decay. The experiences of failure were as diverse as they were unique to each composting system. There were participants who practised hot compost, others cold, and some who operated open-air systems. More still had compost encased in plastic domes with access to the soil at the base. A common failure for the hot composters was identified as inadequate aeration that led to the death of the aerobic bacteria, the key microorganisms hot compost relies on to break down matter. When these died, anaerobic bacteria move in, detectable by a halt in the pile's reduction (indicative of a slowing rate of decomposition) and the release of various smelly gases including rotten egg gas, which was recognised by all as an environmental hazard to be avoided. One gardener, Linda, viewed her composting efforts specifically as a means of carbon sequestering and focused on avoiding such failures for environmental reasons.

Through the gardeners' efforts to repurpose organic waste into something capable of nourishing multiple unknowable others—or the very flows of unbounded life—to hasten transformation into material that was 'good for the garden', there was both an awareness of the multispecies endeavour of compost and respect for the process and the interactions involved. Most composters felt they were not doing enough to support these processes. 'I should do more', or, 'I could try this', were common refrains as evident in Gabriella's response:

> Oh, I've got bins, two bins and two bays and I kind of start with the bins and you know every now and then have a big composting bee and sort of, I don't really like everything. I don't do it kind of regularly or it's, you know,

> properly, but occasionally I'll have a big thing and I'll turn it all and cover it and do it properly and I'll come back. Like yesterday and discover that there's a bit of a compost you know ages later but yeah, I'm a bit kind of random about when they're all full I'll suddenly kind of go, 'Okay, I need to tip them out and turn them'.

Despite these expressions of inadequacy, the majority of those involved in composting were producing material that could be used in their gardens. Comments pointing to their need to do more seem to be generated by a feeling that they were not pulling their weight in these multispecies and multi-material endeavours. The gardeners believed they could always do more, or do things differently, to try and support the networked interactions they relied on to nurture their gardens, and thus part of their food supply. Humans were aware of their reliance on the relationships among others and took seriously their task of attempting to understand (and act accordingly) how best to support these interactions. The ongoing playful tinkering enacted with convivial dignity of these composters attests to their need to adapt to, and work with, the unknowable others and the relations that form among them even when the outcomes were unknown and unable to be controlled.

Encounters with transitioning compost

Talk of evolving 'compost recipes' was one way in which this playful tinkering enacted with convivial dignity was evident. One gardener, Jeff spoke of the challenges of getting the ingredients and quantities right. Another, Harry, noted that composting required more than just the dumping of food waste into a pile, instead requiring intense bodily labour to 'gather the right materials' to nurture the heap. This composter ascribed to what he called a homeopathic approach whereby he and his partner aimed to strengthen the capacity of the compost heap and its microbial presences to do its work. While efforts to 'balance' the ingredients were talked about, this required more that strict adherence to predetermined scientific ratios. While most were familiar with the standard guidelines for compost composition (ideally a carbon to nitrogen ratio of 25–30 to one part) these were seen as both difficult to achieve and only one consideration for getting compost 'going well', as indicated by Jacquie:

> It's feast or famine when it comes to nitrogen and carbon; it's always lopsided. And when it's going well, you know, when it's going well it's never an issue because it just keeps going down, down, down, down.

Effective management of the heap required careful observation and responsive engagement. It couldn't be fudged by applying pre-determined formulaic approaches. In a manner akin to the playful experimentation with meal recipes discussed in earlier chapters, the materiality of the pile was seen to demand particular types of attention and ingredients in response to its transitional actualities and potential. However, in the compost heaps where decay is desired, the bounded material entities of individual foods (or food waste) was not the matter being attuned to. Instead, it was the flows among a vast array of materialities, their entanglements and interactions that were the focus. Compost was understood to constitute forms of togetherness among unknowable entities and transformational relational flows. This togetherness could be supported and enhanced, thanks to, not in spite of, alterity.

The desired outcome of these risky forms of togetherness-in-relation formed through playful tinkering is humus rich, 'life-giving' compost. This material assemblage was valued so highly that for one of our participants, Jenna, it was gifted to others throughout her neighbourhood: 'I mean, I pump out a lot of compost each week. And if I can't use it I'll give it away because people want compost for their garden...'. While Evans (2012, 2014) demonstrates that people are extremely reluctant to 'gift' meal leftovers beyond the immediate family, here compost created from this woman's food waste and other 'waste' matter was freely distributed and in hot demand. Compost was 'pumping' out of her heap. Perhaps it was the transformation of this material—its shift from the original surplus state—that rendered it more readily 'giftable'. This was, however, also a householder who shared food amongst her friends and neighbours. While not considered to be leftovers as such (because it was usually the product of intentional big batch cooking), there was already a system of sharing and a strong social network in place that the compost slotted into. Still, its value as a gift was seen as deriving largely from its propositional nature. The compost had already proved itself capable of becoming something new and this was suggestive of its capacity to go on to contribute to more productive life. It was also something that was too good to waste. The generative potential of the material and is generative relational possibilities motivated the composter to move it along to new sites.

Overall, the decomposition process of compost tended to be experienced through recognition of entangled relations, with a focus on the flows of life rather than the uniquely identifiable bounded nonhuman entities that contributed to its transitions. It was through these relational entanglements that mutual vulnerabilities were exposed and within which the capacity for new forms of togetherness enacted with convivial dignity could be glimpsed. However, for those who identified the use of animals, such as dogs and chickens and worms, as key to their management of food waste, there was a tendency (particularly in relation to the first two mentioned) to focus more heavily on the role individual bodies played in decomposition rather than seeing it as a broad multispecies, relational endeavour.

Human-animal food waste relations

The focus on bounded entities was evident when householders spoke of the likes and dislikes of animals and the efforts they expended to get to know these and cater to them. While, in relation to compost, some noted they avoided adding certain food waste items to it (namely meat, bones and citrus), this was not discussed in reference to likes and dislikes. Compost was consistently spoken about as a collective—a site of multiple togetherness—within which the individuals were not discernible and thus preferences were less detectable as they were more open to shifts and change depending on the relational composition. Animals, however, were seen as in need of being cared for in different ways whereby the human was ultimately responsible for their health. This meant, for example, limiting scraps to maintain a healthy diet for dogs. Human responsiveness to the animals was balanced with existing understandings of how to 'provide good care', which involved attempting to negotiate the responsibilities of pet ownership, health advice and management of their food waste (among multiple other household responsibilities).

Expressions of concern for bounded entities, rather than relational flows, was common in households that used dogs and chickens to manage food waste, but this emphasis was not as clear-cut for those who had worm farms. Throughout this book, I have developed the idea of togetherness-in-relation as key to the narrative notion of convivial dignity, which, as identified in Chapter 2 builds on the concept of togetherness articulated by Abrahamsson and Bertoni (2014) in their encounters with vermicomposting. In their work, these authors explore the effects and affects of relations that can move us beyond concerns with attachment and detachment. We don't have to love or hate, or want to be like each other to be able to live well together. Instead, Abrahammson and Bertoni speak of 'compost politics' that is not simply about attuning to others but also their activities; it is about the doings as much as the material entities themselves. They note that worm bins 'complicate the boundaries between the different transformations that go on in and around the bin, between insides and outsides, and even between humans and nonhumans' (Abrahamsson & Bertoni, 2014, p. 136). This echoes the experiences of the compost and worm farm practices of those contributing to this research, wherein the entanglements of the processes of decay that exceeded the contributions of the bounded entities served to highlight interdependencies and mutual vulnerabilities.

However, the worm relations encountered during research with the Canberra participants were more nuanced and seemed to require much less human intervention (sometimes too little) than experienced by Abrahamsson and Bertoni (2014). Some worm farms thrived with minimal attention with Julie, one householder who felt her compost never quite worked, speaking enthusiastically about the great success of her worm farm:

> I've got these lovely little worms and I always use the worm castings in the vegetable garden; so I sort of make sure that I use that. I feed them and chop it all up for them. And I use the juice on the garden too, for fertiliser.

For her, the worms were felt to be much more vital and productive than the compost heap. Others, however, were in constant danger of forgetting their worms, regularly having to 'replace' those they killed through neglect. These worm killers seemed to enact procedures similar to those that Evan's (2012) identified among households where leftovers were delayed from entering the waste stream by storing in the fridge for later consumption that never occurred. Just as fridges and Tupperware containers become 'coffins of decay' in Evans' work (2012, p. 1132), so too did the plastic encased worm farms in Canberra. Tucked away in garden corners, these wormeries often became sites of mass neglect leading to death, rather than the desired sites of decomposition. This was exemplified by one 'worm parent', Gabriella, who, while pointing out that her chickens were thriving, identified a different set of relations with her worms:

> I've also got a worm farm, I'm a really crap worm parent you know? It's like my third lot of worms, I kind of get them and feed them once and forget about them for a few months and then I'll buy some more. You know, freeze them, fry them [laughs]. So yeah, I have all the best intentions about worm farming and, in fact, yesterday I discovered that there was kind of like one lonely worm in there and I gave it a bit of food but this morning the whole worm farms' frost covered so. . .

All of the worm farmers highlighted their 'good intentions', but these were regularly unfulfilled. Jacquie, who confessed to frequently forgetting about her worms, expressed relief that there is 'no RSPCA for worms' but also outlined how, at times, she had invested a great deal of energy to looking after the worms, recounting a period when she trialled blitzing up scraps in a vitamiser to feed them. However, the majority of worm farmers felt that the technology that encased these decomposers—most commonly opaque black plastic—made it easy to forget about them, particularly when food scraps were already diverted from mainstream waste in other ways.

Where multiple means of managing food waste were used in households, a hierarchy of waste distribution developed. Animal bodies, particularly companion species, were the first to be nourished with food scraps. Pets such as dogs and guineas pigs topped the list, usually followed by chickens, worms and finally the compost heap. In households where interaction with compost was minimal, its vitality was rarely encountered and the more overt liveliness of animal bodies ensured they were nourished first. Multiple means of disposal also spread the vulnerabilities; if one didn't work there was a back-up. However, this also required attunement to multiple more-than-humans as exemplified by the system employed by one householder, Lynn:

> Well, it's something I've thought about consciously, and it's probably been. . . we've got this system of three, as in we have worms, we have a normal compost heap and we have an indoor composting bokashi system. So priority goes. . . when there's waste from preparing food or leftovers, the first thought is 'What will the worms get?' At the moment the worms are slowing down because it's cold, so they're getting less because they're just not getting through it. Then it's 'What can go in the normal compost?' and what we use the bokashi for? So what we use the bokashi for are things that don't go as well in worms. . . for the worms, they don't like citrus peel for

> example, so much. And they don't go as well in the normal compost. So, things like bread crusts and that sort of stuff, so we bokashi those things. So, what we've found is that what it does is it speeds up the eventual composting process, and things that would normally take forever, get a boost from the bokashi system. And we put our bokashi into the compost, rather than putting in any other holes or whatever.
>
> I came across the bokashi system because I was conscious of all this food waste that worms couldn't eat and compost couldn't eat, so it was, I don't know, a win fall thing to find out about it and 'Oh, I like that idea, I'm going to go with that and try it'.
>
> It is. It means... because I guess me personally, I was consciously thinking about well what am I doing with all this food waste that ends up in the bin, surely it can be put back even into our garden. So it's reduced substantially. I'm not entirely sure what to do with meat half the time, sometimes the worms get a bit, sometimes a bit in the bokashi, but I'm not... I haven't fully thought through where I want to be with that.

Domestic management of food waste requires an active decision to invite a multitude of other species into our homes and backyards or balconies. These are species that do not always contain themselves within their dedicated locations and feeding them can lead to the arrival of uninvited and undesirable others. Through navigation of these multispecies landscapes, the composters in this research find themselves more closely intertwined with nonhumans as a result of their embodied practices of managing food waste. In their gardens and their homes, embodied, convivial and responsive interactions with food waste seem to facilitate an openness to the lively materialities of these multiple more-than-humans and the relations that unfold.

Conclusion

This chapter has attempted to demonstrate how interactions with food waste reconfigured through compost or the bodies of animals can develop through forms of playful tinkering that manifest in practices guided by convivial dignity. Compost and decomposition, like play itself, is propositional. These are sites which can support engagement in forms of practical and 'ontological choreography' capable of inciting 'efforts to determine what is this, what can it be, how can we be together?' (Haraway, 2008). This togetherness can be enacted through the formation of intimate attachments but is also often marked by strong desires for detachment. But the potentialities that emanate from recognition of our inescapable togetherness do not have to be driven by sentiments of care and a desire to be close nor do they have to manifest in an oppositional form of 'holding-apart'. Such relational modes may well form, but underpinning these and multiple other modes of encounter is the recognition or sensing of the necessity of our mutual entanglements, or what I refer to as togetherness-in-relation.

Indeed, encounters with compost involves recognition of togetherness-in-relation among human, more-than-human and non-bounded entities and their transformational potential. The shifting relations that make composting, worm farming and backyard chicken keeping possible expose mutual vulnerabilities where risky togetherness is understood to involve interactions with radical alterity and thus unknowability. The receptivity, attunement and responsiveness to these unknowns through playful tinkering enables the development of ways of being and doing that can challenge conceptions of human exceptionalism and are suggestive of a generative form of ethico-political practice in the garden, an approach unsettled by but attuned to, life as flows and interdependencies.

Throughout this chapter and the Waste section more broadly, I identify the cultivation of responsiveness to unknowable others and the relational entanglements that enable enactment of life and matter, through playful tinkering with food surplus and waste, as playing a key role in reducing food waste and supporting a rethinking of the myth of hyper-separation embodied in normative notions of the modernist subject. However, in the following chapter, I scale up my focus beyond the micro-worlds of compost heaps and the lived practices of householders and gardens to explore how effective such relations may be at the level of a corporation, focusing on the global ugly food and food redistribution movements. This stretching of scale leads to encounters with the limits of responsiveness, identifying how, when such relations are enacted without convivial dignity, they are unlikely to be capable of supporting non-anthropocentric modes of being and doing. However, when I return to focus on small-scale actors within these corporations and entities, the overarching anthropocentric narratives of the businesses or NGOs are shown often to be at odds with the recognition of the vitalities of matter and shared relational entanglements sensed by workers and volunteers. As such, I use the following case studies to reinforce the need to focus on the potentially disruptive practices of small-scale initiatives that have the potential to promote less resource-intensive modes of living through recognition of shared togetherness-in-relation. This is, once again, shown to be encouraged by participation in practices marked by playful tinkering and enacted with convivial dignity.

References

Abrahamsson, S., & Bertoni, F. (2014). Compost politics: Experimenting with togetherness in vermicomposting. *Environmental Humanities*, *4*, 125–148.

Bennett, J. (2001). *The enchantments of modern life: Attachments, crossings, and ethics*. Princeton and Oxford: Princeton University Press.

Brice, J. (2014a). Attending to grape vines: Perceptual practices, planty agencies and multiple temporalities in Australian viticulture. *Social & Cultural Geography*, *15*(8), 942–965.

Brice, J. (2014b). Killing in more-than-human spaces: Pasteurisation, fungi, and the metabolic lives of wine. *Environmental Humanities*, *4*(1), 171–194.

De, P., & La Bellacasa, M. (2017). *Matters of care: Speculative ethics in more than human worlds*. Minneapolis, MN: University of Minnesota Press.

Evans, D. (2012). Binning, gifting and recovery: The conduits of disposal in household food consumption. *Environment and Planning D: Society and Space*, *30*(6), 1123–1137.

Evans, D. (2014). *Food waste: Home consumption, material culture and everyday life*. London and New York: Bloomsbury.

Ginn, F. (2014). Sticky lives: Slugs, detachment and more-than-human ethics in the garden. *Transactions of the Institute of British Geographers*, *39*(4), 532–544.

Green, K., & Ginn, F. (2014). The smell of selfless love: Sharing vulnerability with bees in alternative apiculture. *Environmental Humanities*, *4*(1), 149–170.

Haraway, D. (2008). *When species meet*. Minneapolis, MN & London: University of Minnesota Press.

Latour, B. (2004). How to talk about the body? The normative dimension of science studies. *Body & Society*, *10*(2–3), 205–229.

Law, J., & Mol, A. (2006). *Globalisation in practice: On the politics of boiling pigswill*. Draft paper, version of 9th April 2009 Department of Sociology, Lancaster University; Department of Social and Cultural Anthropology, Amsterdam University. Retrieved from http://www.heterogeneities.net/publications/LawMol2006GlobalisationinPractice.pdf

Massumi, B. (2014). *What animals teach us about politics*. Durham and London: Duke University Press.

Morton, T. (2010). *The ecological thought*. Cambridge, MA: Harvard University Press.

Rose, D. B., Cooke, S., & van Dooren, T. (2011). Ravens at play. *Cultural Studies Review*, *17*(2), 326–343.

Tsing, A. (2014a). More-than-human sociality: A call for critical description. In K. Hastrap (Ed.), *Anthropology and nature* (pp. 27–42). New York: Routledge.

Tsing, A. (2014b). Strathern beyond the human: Testimony of a spore. *Theory, Culture & Society*, *31*(2–3), 221–241.

10 Ugly food and food waste redistribution

Introduction

Learning to live with abundance and scarcity will be necessary for survival in our contingent futures. In Chapter 8 we explored how these fluctuations are encountered and managed by householders identifying an openness to being moved by the vitalities of food and particular affordances as the basis for avoiding food waste in domestic settings. In Chapter 9, we once again saw that responsiveness to multi-species and multi-relational unknown others can, through playful tinkering enacted with convivial dignity, be generative of non-anthropocentric ways of being and doing with the world. However, in this chapter I question whether such reactionary approaches are useful at larger scales where waste occurs for different reasons and at different points in food flows. To do this, I set my sights on two key movements associated with food waste, namely the ugly food movement, which draws attention to the waste generated by the imposition of aesthetic standards of food distributors (namely supermarkets), and food redistribution services, which redirect and repurpose excess food primarily from food providers (including supermarkets, residential halls and one-off catered events).

Both of these movements are shown to be responsive to the shifting materialities of foods yet fail to attend to the structural problems that generate the surplus in the first place. While ugly food can indeed reduce farm-gate losses, the propensity to sell this at a cheaper price due to the general abundance of supply can obfuscate the over-production issue and the agri-food industry's warped aesthetic standards. Food redistribution manages surplus, rather than working to promote its reduction. Thus, despite their purported aims, these movements can aid and abet the ongoing generation of waste through their responsive practices. At this scale, responsiveness seems to be ineffective in encouraging the rethinking of anthropocentric norms of human exceptionalism that I contend is foundational to our capacity to live well in times of uncertainty.

This is not to say that ugly food and food redistribution don't make important contributions. Both attend to specific problems in food production and distribution and have the potential to improve the wellbeing of certain sectors of the population including the hungry and farmers previously unable to sell their products. They do also have environmental benefits by diverting large quantities of food from landfill.

DOI: 10.4324/9780429424502-10

For these reasons, I do not want to negate the value of the movements but I do wish to question their role in systemic food waste reduction and broader ecological attunement. Overarchingly, the narratives and practices associated with ugly food and food redistribution are replete with tensions and contradictions. For example, there is too much food in some sites, not enough in others. Some consumers wish to support farmers and challenge aesthetic standards by purchasing ugly food, but the goods are cheaper and producers receive less money for their non-standard crops.

In these encounters, the materiality of foods—its appearance, shape, form and its potential to rapidly transition from something edible to waste—is central to its management. Thus, within these practices there is evidence of attunement and responsiveness to the very materialities of food. However, those generating or supplying the surplus can also be seen to defer responsibility for the 'becomings' of waste to smaller players in the system. This enables the 'big' players such as supermarkets to distance themselves from the impacts of their support for particular forms of production, standards and decision-making that enables a 'forgetting' (Hird, 2013) of waste. My discussion of these two movements aims to show that identification of the vitality of matter and responsiveness to this is not always generative of non-anthropocentric practices, suggesting that a lack of recognition of togetherness-in-relation, and thus absence of enactment of convivial dignity, inhibits the generative potential of responsiveness.

However, when I return my focus to small-scale workers and volunteers within these broader movements, I once again find that an openness to being moved by food tends to encourage engagement in practices of playful tinkering with surplus food or food waste that prise open possibilities of decentering human subjects and supporting more sustainable practices. This prompts the suggestion that practices of playful tinkering may not be able to be scaled up and that the narrative force of anthropocentric approaches within these organisations makes enactment of alternative behaviours both difficult to operationalise and narrativise. This is argued to reinforce the importance of material-semiotic action in our efforts to promote alternative subjects for the Anthropocene capable of nurturing more sustainable living practices.

While the material-semiotic practices identified throughout this book are shown to be difficult to translate to larger scales in the food waste movements explored in this chapter, I do not see this as an impediment that needs to be overcome. In fact, the book's focus on small-scale action is a response to the lack, or failure, of significant shifts and action occurring at the organisational and governmental levels. While we must continue to advocate for change at the macro level, it is at the grassroots everyday level that we find the enactment of alternative modes of being and doing, and it is these that we must seek to amplify to counteract the lack of action at larger scales in order to develop the agility to enable us to adapt to future food uncertainty.

Challenging aesthetic standards with ugly food

My discussion of the 'ugly' food movement attempts to keep the householder firmly in view as we consider what happens, and what might happen, in people's homes when they purchase these foods. This section draws on the encounters with wonky veg in gardens and at AFNs of 38 householders. While the entry of ugly food into mainstream supermarkets can significantly reduce farm-gate waste, I want to explore the impact access to a greater abundance of cheaper food might have on food waste generation within homes. If, as reports suggest, post-consumption waste is the largest contributor to food disposal in the minority world (Gustavsson, Cederberg & Sonesson, 2011; p. v; Parfitt, Barthel & Macnaughton, 2010; p. 3065), then access to cheaper produce may simply defer food entering the waste stream from the farm-gate to the householder. In saying this, I want to make very clear that I wholeheartedly believe that ugly food should be part of mainstream distribution channels. Its presence in these outlets has the potential to draw attention to the rather farcical aesthetic standards imposed by supermarkets. However, I am also suggesting that the ugly food movement in its current guise is unlikely to alter the modes of production and distribution that regularly produce minority-world surplus that eventually becomes post-consumer waste. Therefore, I am questioning its capacity to support more sustainable living behaviours and the development of less resource intensive food systems.

Cheapening veg: questioning the value of ugly food

Mainstream supermarket supply of ugly food is trending around the world, spurred on by the 2015 'Inglorious' food campaign implemented by the French supermarket chain Intermarché (2015). Tied to the European Union's year against food waste, Intermarché's campaign set out to 'rehabilitate and glorify' ugly food. Their campaign advertisements depict wonky fruit and vegetables in all of their wayward glory, accompanied by descriptions such as 'grotesque apple', 'ridiculous potato', 'hideous orange', 'disfigured eggplant' and 'failed lemon' (Intermarché, 2015). Alongside the tongue-in-cheek descriptors are reminders that under deformed exteriors lie fresh, nutritious, tasty food: 'a grotesque apple keeps the doctor away as well' (Intermarché, 2015). The initiative reportedly led to a 24% increase in store traffic (Intermarché, 2015) and attracted global attention prompting the introduction of similar campaigns around the world. In Australia, ugly food subsequently arrived in the two dominant supermarkets that have command of around 70% of the nation's grocery shopping market and also has a presence in a number of smaller supermarkets.

In all the messaging employed in the various ugly food campaigns, non-standard appearance is not only identified as having no impact on the nutritional qualities of the foods but also presented as being of benefit to consumers because these goods are typically cheaper (30% in the case of Intermarché) than their more aesthetically pleasing counterparts. This monetary devaluing of the produce reinforces the notion that 'ugly' (even if only skin-deep) equals 'cheap' when it comes to food. Charging lower prices for ugly fruit and vegetables hides the fact that the same human and nonhuman labour and resources are required to produce, harvest and transport crops, regardless of their appearance. As such, ugly food may well help perpetuate a food system that undervalues food.

Let me slow down a little here. So far, this sounds very much like an anti-ugly food argument and that is not at all my intention. My issue is not with ugly food but with its economic value, the price paid to farmers and costs for consumers. As we have seen throughout this text, the choices made about food in households are driven by a host of complex and sometimes contradictory imperatives. However, at the most basic level we know that people do not buy food with the intention of wasting it as evident in the work of Evans (2012, 2014). Domestic food waste is not simply a symptom of a profligate society. The generation of post-consumer waste is attributable to the amount we buy as well as the various impacts on when, how and why this is or isn't cooked and eaten.

In the previous two chapters I have argued that attunement and responsiveness to the materiality of food is a hallmark of those who minimise their waste production. This manifests in a variety of ways, including creating meals in response to what is available (what looks good in the markets/or in the garden) and what 'needs' using (based on a multi-sensorial assessment of shifting material states) rather relying on a set meal plan and shopping list. This responsiveness is driven by multiple factors ranging from environmental concerns to practices of frugality. As outlined in Chapter 5, frugality differs from thrift (Evans, 2011) insofar as the aim is not necessarily to secure a bargain but to avoid wastage. This can be motivated by economic incentives (limited income or experiences of poverty at previous points in people's lives which, through careful management of goods, they believe they can make sure they will not have to endure again), but in this research, money—even for those who identified as having budgetary constraints—was never the principal driver of these practices. Common to all participants was a valuing of the 'life' left in materials and it was this recognition of 'lively' fruit and veg, and their openness to being moved by this, that prompted waste avoidance. This responsiveness to the materiality of food (which I have shown prompts encounters with mutual vulnerabilities through attunement to matter and its transitions that then support recognition of the inescapability of our togetherness-in-relation) can encourage the development of skills and capacities to respond to change and uncertainty. These have been argued to commonly occur in a playful, tinkering manner.

On the other hand, thrifty behaviours, understood here as being characterised by attempts to secure goods for the lowest prices, do not necessarily promote sustainable use of resources. Instead, the procurement of 'bargains' may actually result in less attention being paid to materiality. Cheaper things tend to be plentiful, readily available and thus easily replaced. They are things that many might feel they do not have to treat with great care. In relation to food, such associations could diminish attunement to the shifting states of food for those who are not already engaged in a multitude of practices to minimise waste. As such, thrifty purchasing may actually dull attunement to foods' vitalities.

The lower price of 'ugly' food could also reinforce imposed aesthetic ideals as the norm. Producers have highlighted their concerns about the lower prices non-standard foods attract while supermarkets maintain that these standards, and their price premium, reflect consumer desires (Richards & Devin, 2016). Supermarkets enforce 'strict private standards' such as the 'Woolworths Quality Assurance' in Australia, or collectively owned private standards such as the British Retail Consortium used by Coles with their own 'bolt-on' additions (Richards & Devin, 2016). Given the presence of a supermarket duopoly in Australia (comprised of Woolworth and Coles), the majority of producers have little choice but to attempt to grow to these requirements.

Despite supermarkets' employment of strategies designed to give the appearance of their commitment to waste reduction, ugly food campaigns and donations to food redistributors simply defer the responsibility for the outcomes of their strict standards to other entities. Supermarket responsiveness to the vitalities of food tends to be constructed within dominant neo-liberal modes of valuing, reinforcing anthropocentric imaginings of humans and our capacities to control resources. This responsiveness lacks recognition of togetherness-in-relation and enactment of convivial dignity. As Richards and Devin write, supermarkets do this by:

> . . .setting cosmetic standards in the procurement of food which results in high level of wastage, not taking ownership of produce that does not meet their own interpretation of the standard, claiming corporate social responsibility kudos for donating to food rescue organisations (while at the same time saving on dumping fees) and differentiating between 'beautiful' and 'ugly' foods—reinforcing difficult-to-attain standards of perfection.
>
> (Richards and Devin 2016)

The imposition of these aesthetic criteria employed by supermarkets has a significant impact on the way food is produced and distributed. However, this tends to be obfuscated by the food waste reduction messages they attach to ugly food initiatives.

Celebrating imperfection in AFNs: playing with variability in uncertain food futures

Let me reemphasise that this is not to say that ugly food is without merit. For some, it can draw attention to the arbitrary nature of the aesthetic standards enacted by supermarkets. However, ugly food was largely not on the radars for the Canberrans involved in this research. They had often heard about it and considered it to be a positive challenge to unnecessary and unrealistic aesthetic standards imposed by supermarkets, but as gardeners and AFN shoppers they were regular consumers of food that didn't accord with supermarket standards. In fact, this variation was commonly discussed as being one of the joys of avoiding supermarkets and added a playful element to their food provisioning behaviours. Many were keen on growing or purchasing unusual foods as exemplified by one interviewee who noted, 'everything I buy at the markets is a little bit odd'. In fact, some took children to the markets with the specific aim of enabling them to encounter the diversity of foods available and the variations in their forms with one identifying this as being an antidote to the 'perfection' encountered in supermarkets. Yet, most of the householders interviewed who lived with children indicated that while younger family members were keen on wonky shapes they were rather suspicious of 'blemished' foods. This may be common among many adults too as a study conducted with 4,214 consumers from five northern European countries (de Hooge et al., 2017) found that while people were open to buying 'sub optimal food' for a discount they had a hierarchy of deviations from normal they would be willing to accept and an associated price reduction in mind: shape variations were okay, blemishes were not.

On the contrary, for the gardeners and AFN shoppers in this research exposure to imperfection and engagement in playful tinkerings with how best to prepare these foods was seen to be important to developing their (and where present, their children's) capacity to adapt and respond to the vagaries of food production. In fact, the appearance of fruit and veg was also often used as a talking point with stallholders where climatic conditions, weather patterns or diseases had particular impacts on the appearance and qualities of produce. For some, this provided a chance to connect to the broader challenges of food production and to attune to the interconnectedness of human and more-than-human elements.

Non-standard foods can prompt playful encounters that critique the aesthetic norms of foods and expose people to the actual variability that can occur in food forms. This may be particularly evident in times of scarcity and abundance. In 2012, in response to 'unseasonal weather' the UK supermarket chain Sainsbury's (followed by others) decided to 'radically change' their 'approach to buying British fruit and vegetables' (Vidal, 2012) involving a relaxation of aesthetic standards. The then Director of Sainsbury's Food stated:

> The unpredictable weather this season has left growers with bumper crops of ugly-looking fruit and vegetables with reported increases in blemishes and scarring, as well as shortages due to later crops. We've committed to make use of all fruit and veg that meets regulation and stands up on taste and hope customers will help us all make the most of the British crop in spite of its sometimes unusual appearance.
>
> (Vidal, 2012, online)

Here we can see a willingness to alter standards in the face of perceived necessity. The limits of human control, coupled with the need for supermarkets and their shoppers to be flexible in response to material realities, are emphasised in this example. However, this is portrayed as a mere blip with suggestions things will soon return to normal. While there is some evidence here of encounters with mutual vulnerabilities and recognition of the inescapability of togetherness-in-relation, this is temporally contingent and the expectation is that human actors will again shortly be in control enabling the standards to return.

Uncertainty around climate and weather patterns is a hallmark of living in the Anthropocene and we need to start now to build the skills and capacities to adapt to these unknown and unknowable contingencies. Engagement with ugly food provides the potential to do just this, but when it is cheaper than its standard counterparts, and represented as something that occurs only in unusual circumstances, this cannot be realised. While providing opportunities for training for uncertainty through encouragement of playful tinkering at the householder level, ad-hoc responsiveness at the scale of the corporation does little to question the privileging of certain aesthetic forms and perpetuates anthropocentric beliefs and behaviours, missing the opportunity to support reconfigurations of human/more-than-human relations that are attuned to, and agile in the face of, increasing uncertainty. Large-scale corporations' deferral of responsibility for avoiding food waste, and missed opportunities to support the development of the skills and capacities needed to respond to our contingent food futures, is also evident in practices of food redistribution. Within this growing movement small-scale, not-for-profit organisations are bearing the burden of the big players' generation of surplus. It is to these concerns that I now turn.

Food redistribution: deferring responsibility for surplus

The 'inglorious' food campaign of Intermarche was a precursor to the French Government's 2016 introduction of legislation banning supermarkets from discarding, or intentionally spoiling, unsold food. The nation's supermarkets are now required to donate excess food to charities for further food redistribution or to ensure it is used as animal food. While on the surface such initiatives appear to be a positive step towards waste reduction and respect for resources, like ugly food initiatives, these requirements enable supermarkets to defer responsibility for their waste to others. Surplus in this system is being repurposed rather than removed. Over-production and over-ordering to ensure choice is available to potential consumers is not addressed. Instead, this surplus becomes the problem of other, often smaller, not-for-profit entities. The extent of the potential burden of this deferral of waste was evident in conversations conducted with eight managers, workers and volunteers at food redistribution services in Canberra. As one noted, any rapid moves to introduce similar legislation in Australia would be 'terrible':

> we couldn't handle it, we don't have the facilities or the assets and the infrastructure to handle it, to cope with that amount of food and we don't have enough places or charities in the A.C.T to take all that food, there's simply not that many.

Food rescue has grown significantly in recent years due to the existence of a food system, and players within it, that produce abundance. The scarcity encountered by the hungry provides a ready-made market for redistribution. The arguments developed in this section are not an attempt to deny that food rescue addresses multiple systemic problems with the food system. However, I do emphasise that these are problems induced by the very systems themselves, highlighting the need to attend to these larger scale issues rather than solely focusing on stopgap responsive redistribution. As I suggested in the case of ugly food, micro-level encounters with rescued food can support playful, responsive engagements with its vitalities, encouraging attunement to matter-in-relation possibly capable of shifting ways of being and doing from anthropocentric modes to more generative alternatives. However, larger players in the system tend to respond to the transitioning of food in mechanistic ways that ensure surplus is moved on, quickly deferring responsibility—and thus preventing opportunities for the development of recognition of togetherness-in-relation and enactment of non-anthropocentric practices marked by convivial dignity.

The affective force of multisensorial encounters with surplus food

The repurposing of food provides a site where there is immediate and ongoing engagement with abundance and scarcity. At the micro-level this involves the intended recipients of the food, but my focus in this section is on those responsible for its collection and redistribution. For these people, the vitality of the produce incites action and induces pleasure and satisfaction. This is immediately obvious when they talk of the 'thrill' of picking-up certain foods and their identification of 'life' left in wilted fruit and veg, as exemplified by Ted, a paid employee:

> Veggies are always a major excitement, getting vegetables, so you bring that back. Getting canned food like baked beans and spaghetti and soups is very much an exciting product that we pick up all the time, it stores well and it's not got so much of a use-by date. We can always do with more vegetables though, that's always number one. Milk is very good, too. When we get milk, we get quite a big batch of it, almost to the point where we store our main ones here but we do need to offload, too. We offload to those other charities or refuges that we go to.

The affective force of these encounters is induced through embodied, visceral experiences. Cleanliness, training, observation and sensorial engagement to assess food safety are central to food rescue practices, as noted by another paid employee, Dan:

> At each pick up our van will arrive, our vans are refrigerated and they have scales in them, they're always kept very clean. We'll arrive at any pick up point, go and talk to the people who are storing the food for our team. We'll then discuss with them what the food is, what type of food is there, if it is prepared food, how long can it be frozen, does it need to be eaten same day, that sort of thing. If it's produce we inspect it as well to make sure there's nothing with any visible signs of decay or it's not passed its use-by date. We inspect the food as well.

> We then separate it into produce, fruit and vegetables things like that, dry goods, meat, prepared meals, dairy, bakery, things like that and we weigh each item and variety, we then put that into our application, the app we have on our phone, that collects all the data on everything we collect. We then close up the doors and drive, we have a set run every day, we visit a certain number of food donors and a certain number of charities so people have some certainty knowing that whatever we get that day we will arrive with...
>
> And then we take it straight as quickly as possible and efficiently as possible pretty much from where we've picked up to the closest charity, that's efficient. We let them know we're coming in most cases or they happen to know anyway and are expecting us. We'll then open up the van and say this is what we have, what can you use? And they will take a variety of food, fruit and vegetables, some bakery, dairy products, a range of food that they know their clients will need. That's pretty much it.

The food is carefully handled, visually inspected and other senses used to engage with its vitality to assess its potential to nourish others. The steps taken in these encounters were regularly linked to a desire to respect the myriad of human and more-than human resources, from soil to water, that produced the food, invoking recognition of togetherness-in-relation that is key to enactment of convivial dignity and which was often responded to and enacted in a playful manner. Indeed, attunement to these relational entanglements directly impacted on the food management practices of the rescuers. Yet the story of food redistribution is not simply one of harmonious human/more-than-human relations.

Overarchingly, encounters with surplus food within redistribution organisations are marked by two somewhat contradictory positions that enliven the tensions of contemporary experiences with abundance and scarcity: firstly, the determination to rescue and redistribute surplus in order to meet the needs of the hungry; and secondly, despair at the generation of such surplus and engagement in efforts to encourage its prevention. While rescuers' responsiveness to the materiality of food has the potential to incite recognition of togetherness-in-relation promoting engagement in playful tinkering guided by convivial dignity to avoid waste and support reconfiguration of human/more-than human relations, this is only possible if the supporting larger-scale infrastructure and affordances are in play. Before returning to the Canberra case study to flesh out these contradictions, I briefly map out the broader food rescue infrastructure and the historical legacies of these tensions.

The changing face of the charitable food sector: critical and advocate perspectives of food rescue

Until recently, food redistribution was dominated by food banks, defined by Lindberg et al. as 'not-for-profits that have major food industry partners' (2015, p. 359). Food Banks store large volumes of non-perishable produce that is distributed to local community or charitable groups that then provide this to those in need, namely 'the most disadvantaged community members, including migrants, people experiencing homelessness and the working poor' (Lindberg et al., 2015, p. 362). This so-called charitable food sector is commonly populated by religious-based groups and other NGOs (Edwards & Mercer, 2013, p. 179). However, growing attention to the rates of fresh food wastage over the last two decades has given rise to new forms of food redistribution—what I, following Lindberg et al. (2015), refer to as food rescue organisations. These are characterised by being not-for-profit groups that collect surplus food, cooked and fresh, from a range of venues such as supermarkets, farmers' markets and residential halls and, on a more ad-hoc basis, from sites where events such as conferences and weddings are held.

Food rescue is regularly identified as being a highly successful waste-avoidance strategy, enabling surplus to assist the less fortunate and fulfill their 'right to food' while reducing organic waste in landfill. Representations of these services draw heavily on rights discourse and narratives of anthropocentric forms of care for those less fortunate and the environment. The notion that food rescue provides a win-win solution has been emphasised since the mid-1990s when food waste rose to the top of the international minority-world waste agenda, becoming key to strategies designed to significantly reduce or divert waste and vital in efforts to promote sustainable futures.

Analyses of food redistribution reflect either an advocate or critical perspective (Vlaholias, Thompson, Every, & Dawson, 2015). The former overtly positions it as part of the moral economy: it is good for businesses who don't have to pay disposal costs, good for the hungry who need the food, and good for the government who benefits from avoiding a formal role in these encounters while being rewarded with a reduction of waste in landfill and contribution towards realisation of its sustainability goals. In Australia, while Vlaholias et al. (2015) contend that key food rescue agencies including OzHarvest, Foodbank, SecondBite, and Fareshare, adopt an advocate perspective, those involved in the sector have been shown to be conflicted by the apparent contradictions in their mandates which requires waste to feed those in need but highlights the need to reduce its generation (Lindberg et al., 2015). Those who point to this tension argue that 'food redistribution highlights the inefficiency of our food system—geared towards overproduction, with food redistribution viewed as a "moral" form of disposal' (Vlaholias et al., 2015, p. 7999). Indeed, Warshawsky, drawing on a political ecology approach in his study of a food redistribution agency in LA, asserts that those involved in rescuing food 'may contribute indirectly to neoliberal governance models when they romanticize the power of local communities, focus on individual responsibility and depoliticize food issues' (Warshawsky, 2015, p. 28). The focus on the fulfillment of rights and the provision of care enacted within these organisations tends to work against the politicization of these broader issues, enabling key players in the food system to defer their responsibility.

Practices of food redistribution have a history of being robustly critiqued (see Poppendieck, 1999) for reasons including: its provision of nutritionally unsound foods; lack of attention to consumer acceptance of food; inefficient operational logistics; the instability of supply and distribution; and, what Vlaholias et al. call 'the costs to human dignity and the social othering that occurs when one receives charity' (2015, p. 8000). While a critical approach has primarily been adopted in relation to food banks, these concerns have been shown to also be problematic in the more recent and rapidly expanding food rescue organisations. While there is some evidence of concern, nutritional needs and acceptability tend to play a minor role in food redistribution. In regards to nutrition, Wilson et al.'s research (2012) on the Melbourne arm of the Australian food rescue organisation, Second Bite, demonstrates that deliveries are largely 'ad hoc' in nature prompting the observation that:

> In order to support the good intentions of these charities and ensure that they are maximally effective in improving the dietary intake and health of the beneficiary groups they service, greater efforts need to be made with more accurate data collection, recording, and analysis procedures informed by evidence-based nutritional guidelines.
>
> (Wilson et al., 2012, p. 251)

The following section explores these issues through the encounters with surplus of the two primary food rescue operations in play in Canberra.

Food rescue in Canberra: the challenges of planning for, and responding to, surplus

Canberra is a considered to be an affluent city. As one volunteer noted in conversation, 'people don't think Canberrans are poor' however, a 2013 report commissioned by Red Cross and Anglicare found that food insecure households were found throughout the city, with people often struggling to access food relief services due to their location and the lack of public transport routes to reach them. Food rescue began in the city in early 2008 as an affiliation between the NSW based national organization OzHarvest and the local not-for-profit organisation Communities at Work (CAW) agency. They were known as the 'Yellow Van' service, with one donated van painted bright yellow. Following the national OzHarvest model, food donors needed to register with the organisation prior to the collection of surplus food that would then be delivered to the closest charity where it would be put to use. However, in 2012, CAW split with OzHarvest, declaring it was time to strike out on their own.

For the following two years, CAW ran the only food redistribution service in Canberra, growing to have three vans in operation. However, in 2014 CAW began to implement significant changes to its mode of operation, becoming a central distribution service and introducing an administration fee for charities. Food was collected each weekday from donors, but instead of being redistributed on the road, it was taken to the organisaton's headquarters in the capital's south. In the majority of cases, CAW took orders from clients and these were prepared for collection. A CAW representative interviewed for this research believes that the introduction of the ordering system significantly reduced the waste generated by the charities they work with. This introduction of a more carefully planned approach to anticipated food needs and food surplus represents one way of responding to the uncertainty frequently identified as a problem within the sector. However, this was only sustainable due to CAW's relationship with a food bank that enabled them to 'order' and 'fill' particular requests most of the time. The fluctuations in the forms of other food rescue involving less predictable surpluses, are not as easily accommodated in this model.

The shift in the CAW operational structure meant they no longer accepted all offers of food. In fact, they started to regularly refuse donations as they had no specified redistribution plan or because the collection of the goods was deemed to be inefficient (for example, there was too little offered or it was too close to its need to be used). These shifts in the mode of operation caused concern among some in the community prompting a number of local charities and food donors to approach OzHarvest to return to the city, which they did in 2014. As an OzHarvest Canberra employee noted, 'We won't say no if the food's ripe for us to collect, we'll take it. And then our aim is to get it out to the people before the end of the day, as much as we can'. The two organisations stress that they are not in competition with each other and that there is plenty of excess to go round and people willing to receive and repurpose it.

Challenging donors' conceptualization of surplus

Food rescue services have only become possible in Australia thanks to the success of Ronni Kahn, the founder of OzHarvest, and a suite of advocates and pro bono lawyers who lobbied for the legal changes necessary to enable people and businesses to donate food they declare in good faith is still edible and has been stored and transported appropriately. Prior to the introduction of the National Good Samaritan Act, and in the ACT, changes to the food safety laws and the Justice and Community Safety legislation, donors could be held legally liable if anyone became ill from consuming their surplus food. However, in 2015 (when these conversations took place), after seven years of food rescue organisations running in Canberra, the redistribution organisations were still encountering businesses that did not believe they could donate their surplus food. As an OzHarvest representative observed:

> The law's now been changed around the country but the understanding of the public hasn't kept up with that, a lot of people still think it's illegal so we have to go in and educate people and in these cases, in the case of supermarkets we take the paperwork we show them it is legal and we say look Government House donate to us, Parliament House, Convention Centre, Aldi, everybody else they donate and not a single problem. And they still go, 'Oh... still a bit worried about it though'.

Not being aware of, or not understanding, the legislation is only one issue that impacts on the amount of food rescued. While between them OzHarvest and CAW have 60–90 regular donors, both indicated that more food could be redistributed if there was a greater commitment from the actual donors themselves. Representatives of both organisations observed that donations were regularly treated as if they were already 'waste', seen as no longer useful to the donors and simply entities that needed to be moved on as quickly as possible. Both groups lamented the amount of time and effort that some donors invested in maintaining the quality of the foods to be collected, as noted by Rebecca, a paid employee:

> There are some donors that are fantastic and everything's really clean and everything's really marked there for you and they've really looked after the produce that they're giving to you. Then there are other donors that you almost think that it's a garbage pile, and so you have to sort through the food that you can actually take, so it's just making sure that your hands are clean and that you're mindful, I guess, of where you are.

This management of donations was seen to largely come down to the actions and attitudes of individual donor managers and how they perceived and valued the act of redistribution and the food itself. A high turnover of staff, particularly at larger supermarkets, was identified as a barrier to building the relationships that would help overcome these issues. However, the rescue organisations also understood that coordinating the surplus for rescue was more time consuming than previous practices of binning food and that these new duties were not always factored into the employees' jobs. Thus, while supermarkets were viewed as willing to support repurposing, extra commitment in terms of time (and thus cost), as Rebecca observes, was not always evident:

> These major food chains, like, the supermarkets, have a huge turnover of store managers and dock hands. If you don't have a good dock hand then you get limited produce. If you don't have a very good store manager who believes in the philosophy of saving food and rescuing food and spending maybe that fraction extra in administration, then you don't get the produce.
>
> So there's always this negotiation with these major chains in the importance of spending just that fraction of extra time to be able to provide us with the produce.
>
> So you could have a very enthusiastic dock hand who is happy to do that and a store manager who insists that the staff find that time to do that, but you can have ones who just won't give an inch, and that's the frustration as well.

These uncertainties mean that the quantities and qualities of donations were regularly unknowable. It also meant that food rescue employees and volunteers had to actively assess the quality of foods in all collections—looking, smelling, touching. While the donors often prioritised other values (namely economic considerations) over the vitalities of foods, the inputs that produced them and their potential to nourish others, the food redistribution workers were required to engage in multi-sensorial, multi-material attunements to these entities and the relations in which they unfold. As such, these visceral encounters with the collected foods entailed exposure to vulnerabilities and recognition of togetherness-in-relation particularly when dealing with poorly managed surplus for donation. For one Dan, this highlighted the need to focus on education, information and transparency around waste:

> I believe that the first thing and the most important thing that needs to happen is education and a real, an honest and real commitment from the food retailers, the large supermarkets, to an honest audit at the end of every day about what's still good and what's not. Don't just dump it.
>
> If that means putting on an extra staff member to really go through and sort and I've got that to charity and that to stock, cattle, whatever, fine. Now it may cost you another wage but you can write that off, if you want, on tax you know. But you're doing such good in the community. They need to get real about it; they need to, because it can save them money.

These comments highlight the complexities of surplus, and the need to be viscerally attuned to the vitalities of food and to respond to recognition of our togetherness-in-relation if maximum amounts of food waste are to be avoided. This does not have to manifest in forms of playful tinkering, but it does require a willingness to shift practices and disrupt accepted narratives of value within these corporations. The becomings of food waste demand attention and this can take time and costs money at the organisational scale. In these experiences with supermarkets we can see that food redistribution groups exert efforts, and express desires, for more engaged attention to surplus to ensure it can be repurposed. However, the mandate of many food rescue organisations also requires them to support initiatives that aim to reduce the initial surplus through a reconfiguration of systems of over production and consumption. The following section explores the tensions apparent between these varied aims.

Uneasy relations: the impacts of abundance and scarcity

Food rescue organisations are often not only focused on redistribution but also often tasked with promoting waste reduction behaviours. This is true for the Canberra arm of OzHarvest, as noted by Dan:

> in a perfect world, we wouldn't need to operate because there'd be no food wasted, you know, the big retailers would be smart enough. But even with all the latest software and the latest technology mistakes are still made you know.

However, further reduction in waste from supermarkets was a concern for the redistribution needs of the other food rescue organisation, with Rebecca observing that:

> ...the kind of issues that we're struggling with at the moment is how we move along when supermarkets are trying to now look themselves at their waste.
>
> I know a lot of supermarkets are going into automatic ordering so it's not reliant on humans now, and so their waste issue is becoming less because it's being automated and it's reducing the amount that we're getting, so it makes it hard in that regard.

Rebecca also noted concerns related to the ugly food movement, observing that access to wonky veg had been important to her organisation's provision of food, making a significant contribution to the food supplied from centralised food banks. She expressed concern that selling ugly food in mainstream supermarkets would make this less affordable to food banks:

> At the moment, your bigger supermarket chains are now starting to sell your deformed potatoes and bananas and all of that kind of thing, so those produce now aren't going to your food banks, making it difficult again for lower income earners to be able to have access to that food.
>
> They're being sold just for slightly less at the supermarkets but still unaffordable for our shoppers but because they are going back into the supermarkets they're not being left for us to pick up through the van or through Food bank.

Issues of abundance and scarcity in relation to quantity and access to food dominate experiences of food redistribution making for uneasy relations. While all organisations express dismay at the volume of food waste generated, their services are reliant on, and responsive too, this. Above we can see that, on the one hand, a 'perfect world' eliminates the need for these organisations while, on the other, the presence of hungry people raises concerns about the loss of surplus. This echoes the findings of Lindberg et al.'s (2015) study of the food rescue organisation, SecondBite, in Melbourne (Australia's second largest city). As exemplified by one of their focus group participants the organisation needed 'waste to feed the needy' and so it 'was not in the organisation's interest to describe this problem or seek to address it, but rather make the most of "surplus" food as a means to support "people in need"' (Lindberg, Lawrence, Gold, & Friel, 2014, p. 1484).Yet, broader advocacy of waste reduction within the systems of production and distribution continues to be evident in the food redistribution movement. This contradiction in aims tends to be attended to by focusing attention on waste reduction at the householder level, rather than at the key corporate players in the food system.

Food rescue and householder waste reduction: valuing the vitality of human and nonhuman inputs

The focus on reducing post-consumer waste is evident in the work of OzHarvest in a myriad of ways from their publication of a cookbook featuring recipes from celebrity chefs on how to use up food that might otherwise rot away, through to their hosting of national events such as Think. Eat.Save. This event ran from 2012 to 2014 in partnership with the United Nations Environment Program (UNEP) and the UN's Food Agriculture Organisation (FAO) Global Initiative on Food Loss and Waste Reduction (SAVE FOOD). It aimed to 'raise awareness of global food loss and waste reduction' by hosting simultaneous serving of meals around the country produced only from rescued food. Top chefs joined the cooking and there were signature events held in capital cities. The use of tired, damaged and ugly food by respected chefs aimed to demonstrate the ongoing value and potential of food that was not particularly aesthetically pleasing. These practices can be seen as supporting engagement in playful tinkering responsive to the relational vitalities of potential food waste. As Dan commented:

> food doesn't have to be beautiful to taste beautiful. So yes, it's at, well that's what I think it says about, at a public level, it says food's precious and the water that goes into making it and the fuel that goes into packaging it and then transporting it, it takes so much, don't just throw it away.

The highlighting of the waste of resources and environmental damage that occurs when food enters the waste stream often manifests within these organisations in ways that position contemporary householders as unthinking profligate consumers simply in need of education. As Dan noted of OzHarvest's upcoming Think.Eat.Save event:

> And we'll have signs and people helping us there to highlight the issue of food waste to say think . . .about what you eat and save, don't buy too much. We'll have a lot of information we can give them just saying you know write a shopping list, budget, don't buy too much, you know. Get some chickens or you know get a compost heap or worm farm, don't waste food.

However, despite this enactment of a deficit approach, the focus on the embodied visceral experiences of cooking and waste management that they encourage hint at the potential of these encounters to generate recognition of togetherness-in-relation capable of supporting the development of non-anthropocentric modes of being and doing.

Playful flexibility: making do in the kitchen

Attention to how to use rescued food, food that was simply a little past its best, or unfamiliar ingredients is something which occurs in both organisations in Canberra through regular cooking classes and through involvement of secondary schools in taking donated food and producing value-added products returned to the food rescue organisations for distribution. In the CAW cooking class for low-income clients observed for this research, the participants could use only donated food. There was always uncertainty about what would be available on the night.

Rather than a hindrance, the flexibility and adaptability this 'unknowability' required of the cooks was seen by the volunteer coordinators of the cooking class to offer the right sort of training for how to make do in everyday life. The ways in which this played out, and the ways in which it was discussed, speak to embodied forms of playful tinkering that provide opportunities to train sensitivities and develop the skills needed to be agile in the face of uncertainty. I recount in full the following discussion with 2 volunteer leaders, Maria and Kim, of the cooking classes to highlight how this unfolds:

MARIA: . . .you're teaching them to use what they've got. They don't have to use expensive products to cook with, you can use your basics and just basically make a meal out of what you have in your cupboard, even if you haven't got a lot you can make a meal out of something you've got.

. . .If you watch all those cooking programs on television you have to have smoked mushrooms, you know, whatever, to make a particular dish. I don't work that way.

KIM: Me neither.

MARIA: We cook what we've got. I go shopping at the markets to buy fresh fruit and vegetables and we eat what's there. And if we don't have it, you don't have it and you substitute and that's the ethos here, which is good.

KIM: Yeah. And if you've got a fridge full of slightly old vegetables, which is often the case here [at the food redistribution agency], you make a pot of soup or you can make a stew or something out of it, it doesn't... you know, there's always something you can make out of something you've got.

This form of responsive making do is encountered, in a playful way, as a challenge. The vitalities of the foods themselves and the capacity for them to be transformed into something delicious and able to nourish and support life is the focus. As another volunteer notes of the involvement of schools in the value-added program, it is about learning 'how you can reinvigorate food that would have otherwise gone to waste'. Being able to do this requires skills and capacities and openness to the vitalities of food and appreciation of the capacities for relational reconfigurations of this matter. Such experiences with rescued food can assist in a heightening of sensitivities attuned to these vitalities. As Rebecca says, 'We will take food that looks—like, your fresh produce that looks like it's almost at its death [laughs] ... because we can still cook with that or we can still do something with it'. While there may be little 'life' left in these foods, they can be used in ways—brought into assemblage relations—with other foods through various skillful processes to nourish bodies. making do is a means of valuing the inputs that produced these foods and recognition of their relational potential.

Attuning and retuning tastes

Trying new foods and the ability to adapt to what is on offer was also shown to help attune people to the seasons and encourage the development of adaptability in the face of uncertainty. CAW also runs a food pantry largely supplied by a food bank where food is heavily discounted and fresh fruit and vegetables are free. While some issues with food rescue could be the lack of choice (particularly with premade goods) and cultural acceptability, the provision of fresh fruit and vegetables and the capacity to develop cooking skills can also provide the basis for playful modes of tinkering that enhances the affordances and capacities to engage in a responsive way with the world. This was evident in a story Brenda, a volunteer, recounted about a recent client:

> Cause what somebody said to us yesterday is... possibly cause they're not having to pay for their fruit and veggies, they get it for free, but they've got more adventurous in what they would try because it's like, 'Well, I'm not going to buy that in the supermarket 'cause it's going to cost me so much; I don't know if I'm going to like it'. But 'cause they can get it here and it's not going to cost them anything, they might try it and, 'Oh, that's not too bad, I might buy that, I might get that again'. So yeah, that's what somebody said to us yesterday, she's been more adventurous on fruit and veggies.

For someone already extremely conscious of price and experiencing the likelihood of hunger, the free fruit and vegetables encourage experimentation. While I argue in relation to ugly food that cheaper food can be problematic, that is when those foods would ordinarily be supplemented with something else. For the food pantry clients, their choice is usually about whether to eat what is available or go without. The cooking classes also encourage playful tinkering. In fact, concerns about lack of responsive cooking skills among clients prompted one recipient charity to ask a food rescue organisation to stop delivering pre-prepared meals (something they were more than happy to do).

Responding to what is on offer, 'reinvigorating foods', and developing the skills of adaptability and flexibility in the kitchen on a domestic scale can all be supported through food rescue programs that foster attunement to the relational materialities of food and encourage resilience in times of uncertainty. Many of those who are recipients of rescued food are attuned to scarcity and uncertainty on a daily basis. The tendency for 'certain' types of foods to be donated (such as bread) also means that those involved in food rescue work, volunteers, paid employees and clients, are exposed to the complexities of managing abundance. These engagements with abundance and scarcity afford the possibilities of making do where mutual vulnerabilities are brought to the fore and participants are responsive to the 'life' left in matter and its potential relational reconfigurations, prompting playful tinkering aimed at avoiding waste and, potentially, laying the groundwork for non-anthropocentric ways of being and doing with the world.

Conclusion

Attunement to the relational vitalities of ugly food and rescued food can assist develop the skills and capacities necessary for living in times of uncertainty.

Multisensorial encounters with food can provide affordances for recognition of the necessity of our togetherness-in-relation where human exceptionalism is questioned. However, throughout this Chapter I have shown that a responsive approach is not enough at all scales to support a reconfiguration of human/more-than-human relations and underpin less resource-intensive lifestyles. While supermarkets and other large produce purveyors are responsive to shifts in the states of food, this typically involves a deferral of the management of this surplus to other smaller players, who are increasingly not-for-profit food rescue organisations. Or, it involves the selling of ugly food at cheaper prices, negating the fact that the produce costs the same as 'normal' goods to produce and transport and reinforcing arbitrary aesthetic standards. Unsurprisingly, conceptions of economic valuing are at the heart of these practices and this privileging works against the realisation of the generative potential of attuning to the becomings-with of waste and the capacity for this to induce recognition of our togetherness-in-relation manifesting in enactment of convivial dignity.

Thus, responsiveness has been shown to not always induce attunement to, and recognition of, the relational vitalities of matter. Furthermore, it does not appear to support systemic changes to the modes of over-production that generate surplus and necessitate the labour intensive activities of food rescue. Scale appears to be an impediment to the utility of responsiveness in realising less resource intensive lifestyles. However, in my discussion of these two movements we have seen that, at the smaller domestic scale, the individuals involved in these practices are able to train their sensitivities to attune to the ways in which food surplus and potential waste bring our mutual vulnerabilities to the fore. This prompts practices of 'making-do' that leads to engagement in playful tinkering with recipes and food plans within which the life of matter is extended through relational reconfigurations.

At this small-scale, responsiveness is shown to enhance recognition of the inescapability of our togetherness-in-relation prompting an avoidance of discarding food and the development of the skills and capacities to respond and adapt to uncertainty. It is just this agility that will be necessary for living in our volatile futures and small-scale practices are the sites where we find the most innovative examples of this. As such, while we continue to see failure at the Government and corporation scales to adequately attend to issues of climate change, everyday lives of everyday people are rich with actions that work towards mitigating the damages wrought by large-scale human actions driven by myths of hyper-separation. Convivial dignity has been presented throughout this text as a way of narrativising—or providing a grammar—to conceptualise the ethico-political basis for these non-anthropocentric ways of being and doing and to enable their amplification. It is the possibilities of this alternative grammar that I now turn to in the concluding chapter of this book. There I consolidate just why such a narrative maneouvre is necessary as well as attempting to map out how it can be enacted and how we can recognize its impacts. Thus, my aim in this final chapter is to confirm that efforts to promote more sustainable living behaviours must involve action on material-semiotic fronts.

References

de Hooge, I., Oostindjerb, M., Aschemann-Witzelc, J., Normannd, A., Mueller Loosee, S., & Lengard Almlig, V. (2017). This apple is too ugly for me!: Consumer preferences for suboptimal food products in the supermarket and at home. *Food Quality and Preference, 56*(A), 80–92.

Edwards, F., & Mercer, D. (2013). Food waste in Australia: The freegan response. *The Sociological Review, 60*(S2), 174–191.

Evans, D. (2011). Thrifty, green or frugal: Reflections on sustainable consumption in a changing economic climate. *Geoforum, 42*(5), 550–557.

Evans, D. (2012). Binning, gifting and recovery: The conduits of disposal in household food consumption. *Environment and Planning D: Society and Space, 30*(6), 1123–1137.

Evans, D. (2014). *Food waste: Home consumption, material culture and everyday life*. London and New York: Bloomsbury.

Gustavsson, J., Cederberg, C., & Sonesson, U. (2011). *Global food losses and food waste*. Paper presented at the Save Food Congress, Düsseldorf, Germany. Retrieved from http://www.madr.ro/docs/ind-alimentara/risipa_alimentara/presentation_food_waste.pdf

Hird, M. J. (2013). Waste, landfills, and an environmental ethics of vulnerability. *Ethics and the Environment, 18*(1), 105–124.

Intermarché. (2015). Inglorious fruits and vegetables. Retrieved from http://itm.marcelww.com/inglorious/

Lindberg, R., Lawrence, M., Gold, L., & Friel, S. (2014). Food rescue – An Australian example. *British Food Journal, 116*(9), 1478–1489.

Lindberg, R., Whelan, J., Lawrence, M., Gold, L., & Friel, S. (2015). Still serving hot soup? Two hundred years of a charitable food sector in Australia: A narrative review. *Australian and New Zealand Journal of Public Health, 39*(4), 358–365.

Parfitt, J., Barthel, M., & Macnaughton, S. (2010). Food waste within food supply chains: Quantification and potential for change to 2050. *Philosophical Transactions of the Royal Society B, 365*(1554), 3065–3081.

Poppendieck, J. (1999). *Sweet charity?: Emergency food and the end of entitlement*. New York: Penguin Putman Books.

Richards, C., & Devin, B. (2016, February 29). Powerful supermarkets push the cost of food waste onto suppliers, charities. *The Conversation*, p. 1.

Vidal, J. (2012, September 28). 'Ugly' fruit and veg make the grade on UK supermarket shelves. *The Guardian*. Retrieved https://www.theguardian.com/environment/2012/sep/27/ugly-fruit-vegetables-supermarkets-harvest

Vlaholias, E., Thompson, K., Every, D., & Dawson, D. (2015). Charity starts . . . at work? Conceptual foundations for research with businesses that donate to food redistribution organisations. *Sustainability, 7*(6), 7997–8021.

Warshawsky, D. N. (2015). The devolution of urban food waste governance: Case study of food rescue in Los Angeles. *Cities, 49*, 26–34.

Wilson, A., Szwed, N., & Renzaho, A. (2012). Developing nutrition guidelines for recycled food to improve food security among homeless, asylum seekers, and refugees in Victoria, Australia. *Journal of Hunger & Environmental Nutrition, 7*(2–3), 239–252.

11 New grammars for the Anthropocene

Playful tinkering with convivial dignity

Introduction

Our climate is changing, hastened by human actions and inducing a future of great uncertainty. Global responses to predictions of a vastly changed planet where both resource scarcity and abundance will make the conventional living practices of those in the minority world untenable, lack the force and breadth necessary to both alter and adequately prepare for the unknown and unknowable outcomes of our actions. At the heart of these problems is the dominance and ongoing perpetuation of anthropocentric narratives that champion human exceptionalism. Representations of humans as distinct from nature infuse our daily lives. Plumwood refers to the subject crafted through these imaginings as the Western 'narrative self' (1996, p. 2) which draws sustenance from old 'forms of humanism' that fail to attend to our relational entanglements. We cannot go on like this. If we push past the insidious common sense nature of these claims and attune our senses to living with the world, the very precariousness of these visions can be glimpsed, touched and felt as a shiver running down our collective spine. It is possible for us to sense the world differently. If we are to survive and live well together in our uncertain futures, then we need alternative ways of being and doing that pick at the frayed threads of these human-centric stories. As Plumwood writes:

> If our species does not survive the ecological crisis, it will probably be due to our failure to imagine and work out new ways to live with the earth, to rework ourselves and our high energy, high consumption, and hyper-instrumental societies adaptively… We will go onwards in a different mode of humanity, or not at all.
>
> (Plumwood, 2007, p. 1)

Through the fieldwork explored in this book, we have glimpsed ways in which everyday people not only can but are reworking their lives and lifestyles around food, from taste to waste, to live differently 'with the earth'. In the minutiae of the daily lives of many of those we have encountered we may be able to discern possible shapes that our necessarily different mode of humanity could take. Of

DOI: 10.4324/9780429424502-11

course, these are grand claims to make about the very small-scale actions we have discussed throughout these pages. You may be thinking that it is a rather long bow to draw between peoples' excitement at seasonal variability in AFN shopping or their responsive composting habits and new modes of humanity. You may be right, but in telling these stories I have attempted to draw attention to the ways embodied, visceral everyday practices rub up against common sense anthropocentric beliefs. In the doings and beings in these farmers' markets, kitchens and gardens, the friction is palpable; these bodies and their encounters with the world do not accord with hyper-separation. The affective force of these experiences of being moved by unknown others and the shifting relations through which we all come to live together prompts stirrings of other possibilities.

In attending to these particular sites, my aim has not been to argue that everyone should be (or could be) growing their own food, shopping at farmers' markets and composting. As I hope I have made clear throughout, I do not see any of these arenas as unproblematic. All are imbued with tensions, particularly in relation to systemic issues of access which, following the Hayes-Conroys, I understand not simply as related to socio-economic and spatial or geographic experiences (as significant as these are) but also 'in affective/emotional terms' as being 'about a whole network or rhizome of forces that influence bodily movement, desire, and drive' (Hayes-Conroy & Hayes-Conroy, 2013, p. 84). I have regularly emphasised that I do not want the arguments in this book to be read as advocating a form of blinkered 'good food politics'. My attention to these arenas of food-work has been impelled by efforts to identify how these encounters unfold, their outcomes and the affordances that support these.

As I have gestured towards the potential for being and doing otherwise in the world is also apparent in other sites, such as amongst former steel workers tinkering with materials to reuse and remake in response to scarcity (Carr & Gibson, 2016). As Carr and Gibson contend, these are '[p]eople who are skilled in dealing with the material world in the face of disruption' (2016, p. 307) and it is just such adaptability and agility that may be critical to our capacity 'to cope with volatile futures' (2016, p. 307). While my research focus could have been elsewhere and otherwise, my attention is drawn to 'play with our food' (Mol, 2008, p. 34) due to the everyday necessity of human contact with food's materiality and the relational flows through which these encounters occur.

If we are to promote and cultivate non-anthropocentric configurations of being and doing, then food provides a global stage for local micro-level practices. Food encounters provide sites where matter, or more precisely matter-in-relation or what I have extended throughout this text to the broader notion of togetherness-in-relation, comes to matter through embodied, visceral, sensorial encounters. However, efforts to challenge old forms of humanity cannot rely on reconceptualising the capacities of matter alone. Anderson (2014) has shown that narratives of human exceptionalism can, in fact, be supported by recognition and responsiveness to vibrant matter. Such capacity to enrol materiality in the service of anthropocentrism demands careful attention. Head et al.'s (2015) work on plantiness is one way in which some forms of human exceptionalism are questioned through the development of alternative notions of subjectivity enacted by nonhumans. Plantiness encapsulates the capacity of plants to 'enact distinctive agencies—sun eating, mobile, communicative and [being] flexibly collective' (Head et al., 2015, p. 410). Herein, there remains a 'holding apart' of humans and plants—they are distinct from one another and act on, and with, the world in divergent ways, yet recognition of planty agentic capacity calls into question not only subjectivity as a purely human concern but also the privileging of human normative frameworks. While Head et al. seem to suggest that such recognition poses little risk to the human subject—we cannot, after all, confuse ourselves with plants—I contend that entering into relations marked by such alterity is inherently risky. But, just as we should not shy away from trouble (Haraway, 2008), nor should we avoid the generative potential of risky encounters.

Risky play: tinkering with alternative conceptions of the Anthropos

The dominance of anthropocentric narratives and the practices these fuel has induced significant changes in our planet. The uncertainty of our futures in a world marked by climate change and its myriad of effects demand that risks be taken. Risk does not have to be entered into out of fear. Instead, as Stengers writes, risk is a 'concrete experience of hope for change' (2002, p. 246). These risky endeavours necessitate encounters with the unknown, and unknowable; engagement with 'strange strangers', Morton's (2010) term to capture the alterity inherent in all life forms—we can never fully know or be ourselves alone as humans and more-than-humans are enlivened through a multiplicity of shifting relational interdependencies. Bounded 'others' are not the only strange strangers, but the very relational configurations and flows that bring these into being and shape interactions also figure here. Throughout this book, I have presented play as a variant of experimentation commonly enacted through a tinkering mode as a potentially generative form of encounter predicated on taking risks.

To be 'in play' necessarily involves exposure of mutual vulnerabilities, inducing a 'condition of receptivity' (Green & Ginn, 2014, p. 152) that encourages responsive interactions prompted by recognition of togetherness-in-relation that is capable of generating practices that could reshape the present with an eye to the future. As Rose et al. write, play requires '[a] daring not to be in complete control, or even to really know the other but to play anyway, ... [and] a sense of humour' (2011, p. 336). Play requires and elicits forms of togetherness and what Massumi (2014), drawing on Bateson calls 'mutual inclusion' in which the possibilities of being other and doing otherwise are captured. This togetherness does not require the formation of attachments. It most certainly does not necessitate cosy, harmonious relationships predicated on love or unidirectional care. Togetherness can occur through detachment. Indeed, forms of holding-apart that permeate encounters with unknowable strange strangers do not have to be barriers to mutual inclusion. But, regardless of the mode of relation that marks these interactions, they are able to encourage recognition of the inescapability of our togetherness-in-relation. This is not always a comfortable acknowledgement. It can lead to tense landscapes, but its recognition puts human exceptionalism under erasure and prompts us to find ways of living together regardless of the challenges this may present.

Play does not seek to flatten out alterity or smooth over difference. It can be enlivened by these very encounters, for play is generative when players are 'moved' by each other. In this way, we can think of play as a form of 'ontological choreography' (Haraway, 2008) that involves 'efforts to determine what is this, what can it be, how can we be together?' (Haraway, 2008). In the embodied enactment of these questions through encounters with the unknown, play can provide opportunities to train our sensitivities to be moved by others and the relations that enable, and are generated by, our togetherness-in-relation. As we have seen throughout this text, play tends to unfold in a tinkering, 'bit by bit' manner that involves adjusting and responding to the material limits and capacities of the matter and relational flows involved in food relations revolving around taste to waste. Through these modes of encounter, the participants have been found to enter into risky relations that can enable the development of the skills and capacities needed to respond to uncertainty.

Playful tinkering opens up spaces of becoming-with manifesting in recognition of the inescapability of togetherness-in-relation but not just with bounded bodies. It provides opportunities to exceed the bodies overtly engaged in the choreography to become more than the sum of their parts while being induced by these visceral encounters. Through these manoeuvres, life, and what is understood to be lively, becomes 'a vector of relation and recombination' conceived of as 'flows of energy and materials' (Brice, 2014, p. 180). The generative potential of play, then, is not something limited only to humans and animals. Plants, through their planty subjectivities (Head et al., 2015) and their capacity to affect and be affected (Spinoza, 1951)—thus to move and be moved—could perhaps also play. So too could flows of energy and material for it is these 'more-than' possibilities, those that exceed the sum of the parts and involve responsive adjustments, that makes playful tinkering so productive. But narrativising these ideas and engaging in discourses able to enact them presents as a significant challenge due to the enduring dominance of anthropocentric narratives.

Narrativising non-anthropocentric subjects and practices

Mol writes that 'while material politics may well involve words, it is not discursive in kind' (2006, p. 17). However, as I have argued throughout this book, efforts to promote more sustainable futures require us to engage in material-semiotic efforts capable of shifting the entrenched notion of human exceptionalism that animates the modern subject that we find at the heart of the Anthropocene. We need alternative ways of nararativising the Anthropos that attempt to speak to the 'more-than-we-can-tell' (Carolan, 2016, p. 147) practices mobilised through encounters with the materialities of food—from taste to waste —and the relational flows of life while also leaving space for the excess of playful tinkering and the recalcitrance of matter. We need ways of talking—or, following the conceptualisation of Goodman (2013, 2016), we need new grammars that are able to attend to and amplify just how we can be and do differently.

In adopting this notion of grammar, I am suggesting that the rules of existing dominant narrative structures in the minority world work against the representation of non-anthropocentric modes of being and doing with the world. While this is normally ascribed to matters' recalcitrance and innate non-narrativisability, I contend that we need to work more diligently at reconfiguring our semiotic approaches to find possible alternatives. This is an urgent task. However, as Mol and Law observe the enduring binary divide in English between active and passive, and master and slave, are key impediments to alternative conceptions of subjects that could prompt the asking of a new, more lively set of questions that displace dominant representations of the human subject as a unilateral actor. These would be questions such as 'What do actors do? How are they creative? How do their underdetermined activities help to create or to destroy? What are the possibilities that they condition?' (Law & Mol, 2008, p. 74). As evident in the fieldwork discussed in this book, modes of doing and being that push at the discursive limits of the modern narrative subject are regularly enacted yet participants commonly struggle to conceptualise these in words. Often this results in them resorting to justifying their practices through more 'logical,' accepted political-economic frameworks, regularly prompting them to make light of their affective responses to matter and relational flows that are evident as they go about their food-related practices. Playful tinkering with semiotic structures may provide a means of generating alternatives.

Playing with semantics: the affective force of convivial dignity

Embodied playful encounters ('nonverbal gestures'), for Massumi, provide the affordances for language. He sees these as embedded within, and foundational to, his animal politics, arguing 'the instinctive acts of animals already include language in potential' (2014, p. 45). Within this framework, language, therefore, not only fails to support myths of human exceptionalism but is a reconfiguration of animal play (and, as the boundaries between plants and animals are hard to fix, perhaps too planty play). As Massumi writes:

> The instinctive usage of language consists in a gestural employment of words as catalyzers of language acts effecting direct transformations-in-place that shake up corporeality and rally appetition, propelling life activity in the direction of transsituational variation.
>
> (2014, p. 45)

Language, we could say, is always 'in play'. The inventiveness and necessary togetherness, or mutual inclusion, of playful tinkering then, should be discernible in language. Playful tinkering, pushing out while reflexively folding back in on self, also exceeds these very movements, opening up spaces for something else. Perhaps this provides somewhere for us to start. In spite of justifiable scepticism about neologism, in the spirit of experimenting with the inventive possibilities of playful tinkering, convivial dignity has been offered throughout the book as a way of narrativising an ethico-political guide for playful tinkering that brings matter firmly into our sensorial contact zone.

Convivial dignity is intended to function as a 'linguistic jolt' (Buck, 2015, p. 370) brimming with affective force that simultaneously attempts to represent, and amplify, moments of attunement and arrest, to expose our mutual vulnerabilities and open us to responsiveness to the unknown and unknowable, to encounter opportunities for training our sensitivities to develop the instinct or 'lived intuition' (Massumi, 2014) vital to living in times of uncertainty. As mapped out in Chapter 2, this incorporates 'response-ability' (Haraway, 2008) and 'learning to be affected' (Latour, 2004) but most overtly manifests in what I have termed togetherness-in-relation. Togetherness-in-relation involves not only being moved by matter but also by the propositional and generative potential of the non-bounded relations that enliven it. As such, it involves a broadening out of what we understand as constituting the social. It speaks to relations of mutual inclusion that are not predicated on assumptions of attachment, detachment nor mobilised by notions of care (which I have argued throughout this text have a tendency to reproduce anthropocentric modes of thinking) but are grounded in necessary togetherness and recognition that survival is induced by and depends on these relational interactions. Thus, convivial dignity is offered as a way of respecting and representing these multifarious forms of togetherness-in-relation. It is an attempt to provide a grammar for talking about our existence that strives to encapsulate the multiplicity of ways that life exceeds the sums of its parts.

Learnings from and with the fieldwork

Convivial dignity invokes an openness to being moved by recognition of the inescapability of togetherness-in-relation and playful tinkering facilitates the modes of encounter that enables the development of the skills and capacities to respond, perhaps eventually intuitively, to these. In the fieldwork explored throughout this text, we have encountered people enacting non-anthropocentric, responsive modes of being and doing in their everyday food-practices, from taste to waste, enlivened through playful tinkering. Convivial dignity has been a hallmark of most of these encounters, even when these have incited frustrations at the realisation of human limits and strong desires to exert human control. I have attempted to identify the key affordances that assist in realising these alternative practices. While these include context specific issues such as past histories, access to food-growing spaces, particular forms of household infrastructure and certain modes of food provisioning, there are also more general affordances that can be discerned.

Most notably, these have been identified as recognition of mutual vulnerabilities heightened through multisensorial encounters with scarcity and abundance (where these are non-life-threatening). The training of sensitivities to these risky encounters and the uncertainties of food flows they draw attention to (through responsive, playful tinkering) is shown to contribute to the development of these affordances. This often leads to engagement in practices of making do where life is recognised as being 'left in' matter which is considered to be open to relational reconfigurations capable of generating new forms of life that exceeds the sum of its part, generative of alternative ontologies and world-making practices. These new ontologies and practices have been shown to be rarely underpinned by an environmental ethic, or mobilised through narratives of care and experiences of attachment. Yet these encounters, which involved recognition of togetherness-in-relation, are brimming with affective force and are capable of inducing the skills and capacities (as well as offering sites of training for) contemporary and future uncertainty.

There is much we can learn from the ontological and world-making maneuvers of these practitioners. However, as Roelvink and Gibson-Graham write, this is '[n]ot learning in the sense of increasing a store of knowledge but in the sense of becoming other, creating connections and encountering possibilities that render us newly constituted beings in a newly constituted world' (2009, p. 322). The necessity of forging modes of togetherness experienced by those in this fieldwork is prompted by recognition (however reluctant) of themselves as matter-in-relation 'moved' by the affective force of others or material flows. The scalar limitations of responsiveness and the affective force of matter, however, were highlighted in the penultimate chapter where I introduced the ugly food and food rescue movements. There we saw that, while at the local, domestic household level, responsiveness can be generative of non-anthropocentric human/more-than-human practices, at the larger scale, responsiveness can prompt practices of deferral whereby recognition of togetherness-in-relation can be avoided.

But this has not been a book about the macro level. While gesturing to the systemic problems embedded in dominant global food systems and contending that these are underpinned and amplified by anthropocentric narratives centred on a modern subject configured through modes of hyper-separation, my focus has been on the small scale; on the everyday domestic practices that, yes, are imbricated and often constrained within these broader socio-economic issues but which are also sites of playful tinkering that are inventive, relational and productive of alternative ways of being and doing. As many of us continue to be let down by the inaction of governments and businesses to take action to mitigate climate change, the generative potential of these small-scale sites are increasingly likely to be the places that promote and provide sustenance for alternative forms of world-making.

However, while these non-anthropocentric beings and doings are embedded in the daily habits and beliefs of the gardeners, shoppers and composters we have met, these individuals often struggled to articulate these ideas. As we have seen, material practices can prove resistant to narrativisation. But, as I have suggested, while our narratives nurture the silent ubiquity of anthropocentrism, the possibilities of play afforded by language—or language in play—have not been fully exploited. If we accept that the gestural mode of mutual inclusion is present in animal play, laying the foundations for human language, we must find ways of speaking that allow for the excess and tensions of togetherness. We need to rally alternative narratives that don't relegate these alternative ontologies and world-making practices only to individual entities in domestic spheres. We need to draw on these to broaden the capacity for efforts to promote more sustainable futures.

The notion of convivial dignity developed here is both ethical and political in its drive. It is concerned with supporting multispecies relational becomings-with that give life to alternative ways of being-doings. It is offered in the spirit of playful tinkering as a means of encouraging further speculative, creative imaginings of how we can live and be differently 'together' in the world. It is only one way we might do this. If we are to reconfigure our 'form of humanity', more of us need to play with material-semiotic possibilities. I hope you take up this invitation.

References

Anderson, K. (2014). Mind over matter? On decentring the human in Human Geography. *Cultural Geographies*, *21*(1), 3–18.

Brice, J. (2014). Attending to grape vines: Perceptual practices, planty agencies and multiple temporalities in Australian viticulture. *Social & Cultural Geography*, *15*(8), 942–965.

Buck, H. J. (2015). On the possibilities of a charming Anthropocene. *Annals of the Association of American Geographers*, *105*(2), 369–377.

Carolan, M. (2016). Adventurous food futures: Knowing about alternatives is not enough, we need to feel them. *Agriculture and Human Values*, *33*(1), 141–152.

Carr, C., & Gibson, C. (2016). Geographies of making: Rethinking materials and skills for volatile futures. *Progress in Human Geography*, *40*(3), 297–315.

Goodman, M. K. (2013). Grammars of the Anthropocene. *Dialogues in Human Geography*, *3*(3), 311–315.

Goodman, M. K. (2016). Food geographies I: Relational foodscapes and the busy-ness of being more-than-food. *Progress in Human Geography*, *40*(2), 257–266.

Green, K., & Ginn, F. (2014). The smell of selfless love: Sharing vulnerability with bees in alternative apiculture. *Environmental Humanities*, *4*(1), 149–170.

Haraway, D. (2008). *When species meet*. Minneapolis, MN and London: University of Minnesota Press.

Hayes-Conroy, J., & Hayes-Conroy, A. (2013). Veggies and visceralities: A political ecology of food and feeling. *Emotion, Space and Society*, *6*, 81–90.

Head, L., Atchison, J., & Phillips, C. (2015). The distinctive capacities of plants: Re-thinking difference via invasive species. *Transactions of the Institute of British Geographers*, *40*(3), 399–413.

Latour, B. (2004). How to talk about the body? The normative dimension of science studies. *Body & Society*, *10*(2–3), 205–229.

Law, J., & Mol, A. (2006). Globalisation in practice: On the politics of boiling pigswill. Draft paper, version of 9th April 2009. Department of Sociology, Lancaster University; Department of Social and Cultural Anthropology, Amsterdam University. Retrieved from http://www.heterogeneities.net/publications/LawMol2006GlobalisationinPractice.pdf

Law, J., & Mol, A. (2008). The actor-enacted: Cumbrian sheep in 2001. In C. Knappett & L. Malafouris (Eds.), *Material agency: Towards a non-anthropocentric approach* (pp. 57–78). Dusseldorf: Springer.

Massumi, B. (2014). *What animals teach us about politics*. Durham, NC and London: Duke University Press.

Mol, A. (2008). I eat an apple. On theorizing subjectivities. *Subjectivity, 22*(1), 28–37.

Morton, T. (2010). *The ecological thought*. Cambridge, MA: Harvard University Press.

Plumwood, V. (1996). Being prey. *Terra Nova, 1*(3), 33–44.

Plumwood, V. (2007). A review of Deborah Bird Roses's 'reports from a wild country: Ethics for decolonisation'. *Australian Humanities Review*, 42. Retrieved from http://australianhumanitiesreview.org/2007/08/01/a-review-of-deborah-bird-roses-reports-from-a-wild-country-ethics-for-decolonisation/

Roelvink, G., & Gibson-Graham, J. K. (2009). A postcapitalist politics of dwelling: Ecological humanities and community economies in conversation. *Australian Humanities Review*, (46), 145–158.

Rose, D. B., Cooke, S., & Van Dooren, T. (2011). Ravens at play. *Cultural Studies Review, 17*(2), 326–343.

Spinoza, B. D. (1951). *Ethics Part III. On the origin and nature of the emotions*. (R. H. M. Elwes, Trans.). New York: Dover Publications.

Zournazi, M., & Stengers, I. (2002). A 'cosmo-politics' – Risk, hope, change – with Isabelle Stengers. In Zournazi, M., *Hope: New philosophies for change* (pp. 244–273). Annandale, NSW: Pluto Press Australia.

Index

For Product Safety Concerns and Information please contact our EU representative GPSR@taylorandfrancis.com
Taylor & Francis Verlag GmbH, Kaufingerstraße 24, 80331 München, Germany

www.ingramcontent.com/pod-product-compliance
Ingram Content Group UK Ltd.
Pitfield, Milton Keynes, MK11 3LW, UK
UKHW022316250926
475702UK00046B/865

* 9 7 8 1 4 7 2 4 8 7 5 4 4 *